Einstein vs. Bohr_ Quantum Clash

Einstein vs. Bohr_ Quantum Clash

Rafeal Mechlore

UNIEK ENTERPRISES

CONTENTS

INDEX

Introduction:

1. Brief overview of the historical context in the early 20th century.
2. Introduction to the central figures, Albert Einstein and Niels Bohr.
3. Setting the stage for the profound impact of their quantum physics debates.

Chapter 1: The Birth of Quantum Physics
1.1A look at the development of quantum physics and the key pioneers like Max Planck.
1.2Explanation of Planck's quantum hypothesis and its implications.
1.3Early contributions of Einstein and Bohr to quantum theory.

Chapter 2: Einstein's Skepticism
2.1Einstein's growing skepticism toward the emerging quantum theory.
2.2Detailed exploration of his critique of probabilistic and statistical interpretations.
2.3His famous quote, "God does not play dice with the universe."

Chapter 3: Bohr's Copenhagen Interpretation
3.1Introduction to Niels Bohr's Copenhagen interpretation of quantum mechanics.
3.2Explanation of the principles and key ideas behind his interpretation.
3.3The reception and controversies surrounding Bohr's interpretation.

Chapter 4: Quantum Experiments and Paradoxes
4.1Discussion of experimental evidence and paradoxes that challenged classical physics.
4.2The double-slit experiment, wave-particle duality, and the Uncertainty Principle.
4.3How these experiments contributed to the clash between Einstein and Bohr.

Chapter 5: The Solvay Conferences

5.1Exploration of the famous Solvay Conferences where Einstein and Bohr debated.

5.2Key moments and debates from the conferences, including their disagreements.

5.3The influence of these conferences on the development of quantum physics.

Chapter 6: Einstein's Thought Experiments

6.1In-depth examination of Einstein's famous thought experiments.

6.2The EPR paradox and its challenge to quantum entanglement.

6.3Einstein's attempts to demonstrate the incompleteness of quantum mechanics.

Chapter 7: Bohr's Response

7.1Bohr's responses to Einstein's thought experiments and critiques.

7.2His defense of the Copenhagen interpretation and quantum entanglement.

7.3The evolution of Bohr's ideas in response to Einstein's challenges.

Chapter 8: Legacy and Impact

8.1Discussion of the lasting impact of the Einstein-Bohr debates on physics.

8.2How their clash influenced the development of quantum physics and philosophy.

8.3The broader implications of their rivalry for science and philosophy.

Chapter 9: Beyond Einstein and Bohr

9.1A look at the subsequent developments in quantum physics.

9.2The contributions of other physicists in shaping the field.

9.3The gradual acceptance of the probabilistic nature of quantum mechanics.

INTRODUCTION

In the chronicles of logical history, barely any conflicts have been as mentally arresting and significantly important as the skirmish of thoughts that unfurled between two of the best personalities of the twentieth hundred years: Albert Einstein and Niels Bohr. This scholarly duel, which worked out on the phase of quantum material science, tested the actual groundworks of how we might interpret the actual universe. Their furious discussions, directed through letters, papers, and in-person conversations, uncovered the profound philosophical gap that had arisen in the beginning field of quantum mechanics.

The scenery for this conflict was the mid twentieth 100 years, a period of significant logical upset. Traditional material science, which had ruled for a really long time, was confronting its Waterloo notwithstanding confounding peculiarities at the nuclear and subatomic scale. The world was entering the quantum period, where particles acted in manners that challenged good judgment and appeared to taunt the deterministic request of old style physical science. This time of logical disturbance birthed quantum mechanics, a system that tried to depict the way of behaving of particles at the littlest sizes of presence.

At the core of this upheaval stood two transcending figures. Albert Einstein, prestigious for his hypothesis of relativity and his earth shattering work on the photoelectric impact, was a resolute backer for a deterministic perspective on the universe. He was a firm devotee to the possibility that "God doesn't play dice with the universe," an expression that epitomized his dismissal of the probabilistic, indeterministic nature of quantum hypothesis. As an unmistakable difference, Niels Bohr, a Danish physicist and the planner of the Bohr model of the molecule, supported the arising quantum mechanics and its probabilistic translation. He contended that vulnerability and complementarity were key elements of the quantum world.

The conflict among Einstein and Bohr was not only a logical debate; it was a clash of perspectives. On one side was Einstein, the supporter of a deterministic, requested universe represented by regulations ready to be found. On the other was Bohr, the defender of a probabilistic, innately questionable universe where perception and estimation assumed a focal part in characterizing reality.

In this investigation of the Einstein-Bohr quantum conflict, we will dig into the focal inquiries, discussions, and outcomes of their scholarly duel. We will travel through the early improvements of quantum mechanics, where trailblazers like Max Planck, Werner

Heisenberg, and Erwin Schrödinger set up for the confrontation. We will look at the key discussions that unfurled among Einstein and Bohr, including the renowned EPR oddity and Bohr's reaction. We will likewise investigate how this conflict of thoughts resonated through mainstream researchers and affected the advancement of quantum material science as far as we might be concerned today.

Past the domain of science, the Einstein-Bohr quantum conflict had significant ramifications for how we might interpret the idea of the real world, the constraints of human information, and the connection among science and reasoning. It tested our instinctive ideas of circumstances and logical results, determinism, and objectivity, compelling us to wrestle with the abnormal and irrational universe of quantum peculiarities.

As we set out on this excursion through the quantum conflict of Einstein and Bohr, we will observer the conflict of two splendid personalities, each determined by an energetic quest for truth and a craving to disentangle the secrets of the universe. Their discussions, in some cases rancorous, once in a while significant, keep on resounding in the lobbies of science and reasoning, filling in as a demonstration of the getting through force of scholarly request and the mission to grasp the central idea of the real world.

1. **Brief overview of the historical context in the early 20th century.**
   The mid twentieth century was a period of significant change and commotion across the globe. It was a time set apart by critical verifiable occasions, social changes, logical unrests, and political changes that would shape the direction of mankind's set of experiences. This short outline will give knowledge into the verifiable setting of the mid twentieth hundred years, making way for the turns of events and difficulties that described this critical period.
   The Finish of the Beauty Époque
   The mid twentieth century rose up out of the last part of the Beauty Époque, a time of relative harmony and success in Europe that traversed from the late nineteenth hundred years to the episode of The Second Great War in 1914. This period was described by monetary development, social advancement, and a feeling of hopefulness. Notwithstanding, underneath the surface, strains were stewing, energized by supreme competitions, patriotism, and a quickly changing worldwide scene.
   The Second Great War: The Incomparable Conflict

One of the characterizing occasions of the mid twentieth century was The Second Great War, otherwise called the Incomparable Conflict. The contention, which endured from 1914 to 1918, involved large numbers of the world's significant powers and brought

about phenomenal obliteration and death toll. The conflict had extensive results, including the breakdown of realms, the redrawing of public boundaries, and the significant disappointment of an age that had seen the revulsions of present day fighting.

The Russian Insurgency

In 1917, the Russian Transformation shook the world as the Trotskyites, drove by Vladimir Lenin, toppled the Temporary Government, laying out a socialist system in Russia. This occasion denoted the start of the spread of socialist belief system and the possible arrangement of the Soviet Association. The Russian Upheaval significantly affected international relations, lighting fears of an overall socialist insurgency and adding to the strains of the Virus War period.

The Interwar Time frame

Following the finish of The Second Great War, the world entered a turbulent interwar period described by financial insecurity and political disturbance. The Settlement of Versailles, which formally finished the conflict, forced correctional measures on Germany and added to financial difficulties that would later assume a part in the ascent of Adolf Hitler and the flare-up of The Second Great War.

The Thundering Twenties

In spite of the difficulties of the interwar period, the 1920s saw a social renaissance in many regions of the planet, especially in the US. The "Thundering Twenties" saw a thriving of craftsmanship, music, writing, and social change. It was a period of jazz, flappers, and the preclusion of liquor, and it laid the preparation for the resulting social movements of the twentieth 100 years.

The Economic crisis of the early 20s

The positive thinking of the 1920s came to a crashing stop with the beginning of the Economic crisis of the early 20s in 1929. The financial exchange crash of that year set off a worldwide monetary emergency that prompted boundless joblessness, destitution, and social distress. The Economic crisis of the early 20s had significant political outcomes, as it made ready for the ascent of dictator systems in different nations, including Nazi Germany.

Ascent of Tyranny

The mid twentieth century saw the ascent of authoritarian systems in Europe and then some. Adolf Hitler's Nazi Germany, Benito Mussolini's fundamentalist Italy, and Joseph Stalin's Soviet Association were among the most remarkable models. These systems

smothered contradict, controlled each part of society, and left on forceful international strategies, at last prompting the flare-up of The Second Great War.

Propels in Innovation and Science

The mid twentieth century was likewise a period of striking innovative and logical advancement. Developments like the car, the plane, and the phone reformed day to day existence. In the domain of science, Albert Einstein's hypothesis of relativity and Max Planck's quantum hypothesis reshaped how we might interpret the actual world. These logical leap forwards had broad ramifications for the advancement of present day material science and innovation.

Social and Social Developments

All through the mid twentieth hundred years, different social and social developments tested customary standards and values. The suffragette development battled for ladies' more right than wrong to cast a ballot, prompting tremendous changes in orientation jobs and valuable open doors. The Harlem Renaissance observed African American culture and inventiveness, adding to the continuous battle for social liberties. In the mean time, vanguard specialists and essayists explored different avenues regarding new types of articulation, pushing the limits of imagination.

The Way to The Second Great War

The international strains and annoying issues left directly following The Second Great War set up for The Second Great War. The forceful expansionism of Nazi Germany, the militarization of Japan, and the settlement strategies of Western popular governments generally assumed a part in the flare-up of the conflict in 1939. The Second Great War would proceed to turn into the deadliest struggle in mankind's set of experiences, reshaping the worldwide request and prompting the rise of the US and the Soviet Association as superpowers.

2. **Introduction to the central figures, Albert Einstein and Niels Bohr.**

   Albert Einstein and Niels Bohr, two goliaths of twentieth century physical science, are focal figures throughout the entire existence of science. Their weighty commitments to the fields of hypothetical physical science and quantum mechanics changed how we might interpret the central regulations overseeing the universe. This presentation will give an outline of the lives, accomplishments, and differentiating points of view of these two wonderful people, making way for a more profound investigation of their vital jobs in the logical talk of their time.

   Albert Einstein: The Notorious Virtuoso

   Albert Einstein, frequently viewed as the prime example of the cutting edge logical virtuoso, was brought into the world on Walk 14, 1879, in Ulm, Germany. His initial life was set apart by a gifted interest and an unquenchable hunger for learning. In spite of the fact that he confronted scholarly difficulties in his childhood, Einstein's scholarly ability ultimately became obvious, driving him to concentrate on physical science and math at the Polytechnic Establishment in Zurich, Switzerland.

Einstein's annus mirabilis, or "supernatural occurrence year," happened in 1905 when he distributed four pivotal papers. Among these papers was his hypothesis of exceptional relativity, which upset how we might interpret space, time, and the connection among issue and energy. The renowned condition $E=mc^2$, which epitomizes the equality of mass and energy, is maybe the most notorious aftereffect of this hypothesis. Exceptional relativity broke traditional thoughts of outright existence and established the groundwork for current physical science.

In 1915, Einstein presented his hypothesis of general relativity, which depicted gravity as the ebb and flow of spacetime brought about by enormous items. This hypothesis anticipated peculiarities, for example, the twisting of light by gravity, an expectation affirmed during the 1919 sun oriented overshadow, catapulting Einstein to worldwide notoriety.

In any case, it was Einstein's work on the photoelectric impact that acquired him the Nobel Prize in Physical science in 1921. This peculiarity, wherein light striking a material surface launches electrons, was instrumental in laying out the possibility of quantization of energy and the double idea of light as the two waves and particles.

In spite of his various notable commitments, Einstein turned out to be progressively disappointed with the arising field of quantum mechanics. He broadly announced, "God doesn't play dice with the universe," communicating his distress with the probabilistic, indeterministic nature of quantum hypothesis. This doubt set up for his scholarly conflict with Niels Bohr.

Niels Bohr: The Modeler of Quantum Mechanics

Niels Bohr, brought into the world on October 7, 1885, in Copenhagen, Denmark, arose as the main figure in the improvement of quantum mechanics. His initial schooling in material science took him to the College of Copenhagen, where he would later turn into a teacher. Bohr's impact stretched out a long ways past his examination; he assumed a critical part in encouraging a cooperative climate for maturing physicists in Copenhagen.

In 1913, Bohr presented the Bohr model of the molecule, a weighty structure that consolidated traditional and quantum standards. This model effectively made sense of the otherworldly lines of hydrogen and denoted a critical stage towards figuring out the way of behaving of electrons inside molecules. It presented quantized energy levels for electrons and established the groundwork for the quantum insurgency.

One of Bohr's most critical commitments to physical science was his improvement of the standard of complementarity. This idea, which arose during conversations with Werner Heisenberg, placed that particles could display both wave-like and molecule like properties, contingent upon how they were noticed. Complementarity tested the idea of a total, objective portrayal of actual reality and turned into a foundation of quantum hypothesis.

Bohr's work on complementarity and his Copenhagen translation of quantum mechanics contended that the principal idea of the quantum world was innately probabilistic and that estimation assumed a focal part in characterizing the result of quantum occasions. This perspective straightforwardly gone against Einstein's determinism, making way for their scholarly conflict.

The Conflict of Viewpoints

Einstein and Bohr's contrasting perspectives on quantum mechanics and the idea of reality reached a crucial stage in a progression of popular discussions, letters, and conversations that spread over quite a few years. The most striking of these trades was the Einstein-Podolsky-Rosen (EPR) oddity, a psychological study proposed by Einstein, Boris Podolsky, and Nathan Rosen in 1935. The EPR Catch 22 tried to feature what Einstein saw as the "inadequacy" of quantum mechanics, proposing that it considered the chance of quicker than-light correspondence, a thought that disregarded the standards of relativity.

Bohr, in his reaction to the EPR oddity, safeguarded the probabilistic idea of quantum mechanics and contended that the mystery was an outcome of traditional, deterministic reasoning. He kept up with that no "covered up factors" could give a total portrayal of quantum frameworks and that the demonstration of estimation itself assumed a major part in characterizing the condition of a quantum framework.

These discussions, while set apart by profound scholarly regard among Einstein and Bohr, highlighted the significant philosophical split between them. Einstein stayed a relentless supporter for a deterministic universe represented by stowed away factors, while Bohr advocated the characteristic probabilistic nature of quantum peculiarities and the need to embrace the job of estimation and perception.

Inheritance and Effect

The Einstein-Bohr discusses made a permanent imprint on the historical backdrop of science, impacting the improvement of quantum mechanics and our comprehension of the major idea of the real world. While Einstein's complaints didn't topple the common perspective on quantum mechanics, they prodded further investigation into the translation and underpinnings of the hypothesis.

Both Einstein and Bohr proceeded to make critical commitments to science and kept on rousing ages of physicists. Einstein's later work centered around brought together field hypotheses, trying to bring together the powers of nature, however he eventually didn't accomplish this objective. Bohr, then again, kept on forming the field of nuclear and atomic material science and turned into a compelling figure in the global academic local area.

3. **Setting the stage for the profound impact of their quantum physics debates.**

The significant effect of the discussions between Albert Einstein and Niels Bohr on the idea of quantum material science can't be completely valued without first comprehension the verifiable and logical setting in which these conversations unfurled. The mid twentieth century was a period of colossal commotion in the realm of physical science, as old style hypotheses gave way to the dazing and irrational universe of quantum mechanics. In this account, we will investigate how the stage was set for the Einstein-Bohr discusses, featuring the critical turns of events and difficulties that prepared for their significant effect on the field of quantum material science.

The Introduction of Quantum Mechanics

The underlying foundations of quantum mechanics can be followed back to the late nineteenth century when physicists were wrestling with confounding peculiarities at the nuclear and subatomic levels. Max Planck's progressive thought, introduced in 1900, that energy is quantized in discrete units (quanta), denoted the introduction of quantum hypothesis. Planck's work gave an incomplete answer for the issue of dark body radiation, however it likewise tested the old style comprehension of energy as a persistent and boundlessly distinct amount.

An essential second came in 1905 when Albert Einstein distributed his notable paper on the photoelectric impact. This peculiarity, where light striking a material surface launches electrons, couldn't be made sense of by old style physical science. Einstein's clarification conjured the possibility that light comprised of discrete parcels of energy, later called photons, and that the energy of these photons relied upon their recurrence. This work affirmed Planck's hypothesis as well as established the groundwork for the idea of wave-molecule duality, a focal principle of quantum mechanics.

Wave-Molecule Duality and the Bohr Model

As the mid twentieth century advanced, the wave-molecule duality of particles like electrons turned out to be progressively obvious. This duality implied that particles, for example, electrons showed both wave-like and molecule like properties, contingent upon the exploratory setting. While this idea tested traditional instincts, it was a pivotal move toward grasping the quantum world.

In 1913, Niels Bohr presented the Bohr model of the particle, an original forward leap in nuclear physical science. Bohr's model effectively made sense of the ghastly lines of hydrogen by quantizing the precise force of electrons and recommending that they involved explicit, discrete energy levels inside an iota. This model overcame any barrier between traditional physical science and quantum mechanics, giving a system to depict the way of behaving of electrons in particles.

Nonetheless, the Bohr model additionally brought up new issues about the idea of these electron circles and the strength of particles, making way for additional advancements in quantum hypothesis.

The Heisenberg Vulnerability Standard and Quantum Mechanics

In 1925, Werner Heisenberg planned the vulnerability guideline, a fundamental idea in quantum mechanics. Heisenberg's standard placed that there are inborn cutoff points to our capacity to know the exact position and force of a molecule at the same time. This standard broke the idea of deterministic consistency, recommending that at the quantum level, certain properties of particles are in a general sense unsure.

Heisenberg's work, alongside that of Erwin Schrödinger and others, prompted the improvement of an exhaustive numerical structure for quantum mechanics. Schrödinger's wave condition, distributed in 1926, gave a numerical portrayal of how quantum wavefunctions develop over the long run, foreseeing the likelihood dispersions of particles in different quantum states.

Quantum mechanics, as figured out by Heisenberg, Schrödinger, and others, offered incredible assets for understanding and anticipating the way of behaving of particles at the nuclear and subatomic scales. Be that as it may, it likewise acquainted an essential test with old style instincts: the probabilistic and indeterministic nature of quantum occasions.

The Einstein-Bohr Discussion: A Conflict of Perspectives

In the midst of the scenery of these extraordinary improvements in quantum mechanics, Albert Einstein and Niels Bohr arose as two of the most unmistakable figures in the field, each with a remarkable point of view on the idea of the real world.

Einstein, known for his hypothesis of relativity and work on the photoelectric impact, was a steadfast supporter for a deterministic universe. He accepted that the evident arbitrariness of quantum mechanics covered fundamental "stowed away factors" that, whenever found, would reestablish determinism to the quantum world. Einstein broadly communicated his wariness by expressing, "God doesn't play dice with the universe."

As an unmistakable difference, Niels Bohr was a defender of the Copenhagen understanding of quantum mechanics, which embraced the probabilistic and indeterministic nature of the quantum domain. Bohr contended that the actual demonstration of estimation and perception characterized the properties of quantum particles, and that any endeavor to reveal stowed away factors was a pointless undertaking.

The conflict among Einstein and Bohr was not simply a logical question; it addressed a significant conflict of perspectives. It pitted Einstein's confidence in a deterministic, requested universe against Bohr's acknowledgment of the characteristic vulnerability and complementarity of the quantum world.

The EPR Oddity and Bohr's Reaction

The Einstein-Bohr banter reached a crucial stage with the detailing of the Einstein-Podolsky-Rosen (EPR) oddity in 1935. In this psychological test, Einstein, alongside Boris Podolsky and Nathan Rosen, proposed a situation where two snared particles, when estimated, would immediately influence each other's properties, in any event, when isolated by significant stretches. This evident "creepy activity a ways off" appeared to disregard the standards of relativity and proposed that quantum mechanics was inadequate.

Bohr, in his reaction to the EPR oddity, contended that it was a result of Einstein's emphasis on secret factors and determinism. He kept up with that the Catch 22 was a result of old style thinking and that the essential idea of the quantum world was intrinsically probabilistic. Bohr's reaction built up the idea that quantum mechanics, as it stood, offered a total and precise depiction of quantum peculiarities.

The Significant Effect

The Einstein-Bohr discusses, while some of the time petulant, significantly affected the turn of events and understanding of quantum mechanics. Their conversations

enlightened the profound philosophical inquiries intrinsic to the hypothesis, including the idea of the real world, the job of perception, and the restrictions of human information.

These discussions additionally prodded further exploration and trial and error, adding to how we might interpret quantum peculiarities. While Einstein's mission for buried factors didn't find support in resulting tests, Bohr's probabilistic understanding of quantum mechanics turned out to be generally acknowledged. The Copenhagen understanding, with its hug of complementarity and vulnerability, turned into the predominant perspective on quantum hypothesis.

Also, the Einstein-Bohr discusses provoked researchers and thinkers to wrestle with the significant ramifications of quantum mechanics for how we might interpret the universe. They highlighted the constraints of traditional instincts even with quantum reality and keep on rousing conversations on the principal idea of the actual world.

Chapter 1

The Birth of Quantum Physics

The introduction of quantum physical science denoted a fantastic upset in the realm of science. Arising in the mid twentieth hundred years, this weighty field broke old style ideal models, presented the idea of quantization, and changed how we might interpret the basic regulations overseeing the universe. This broad investigation will dive into the verifiable, logical, and philosophical settings that established the groundwork for quantum physical science, following its advancement from the early improvements in nuclear hypothesis to the historic work of visionaries like Max Planck, Albert Einstein, Niels Bohr, Werner Heisenberg, and Erwin Schrödinger. It will likewise enlighten the significant effect of quantum material science on our advanced world, from the improvement of state of the art innovations to its profound ramifications for the idea of the real world.

1. **The Preface to Quantum Physical science: Early Nuclear Hypothesis**
1. **Nuclear Theories:**
   The old Greeks, including Democritus, guessed about the presence of unbreakable particles, or molecules, as the central

structure blocks of issue.

John Dalton's nuclear hypothesis in the mid nineteenth century laid the foundation for the idea of discrete, resolute molecules.

2. **The Improvement of Nuclear Models:**

In the late nineteenth hundred years, researchers like J.J. Thomson and Ernest Rutherford took critical steps in understanding nuclear design through explores different avenues regarding cathode beams and alpha particles.

Rutherford's model, with a focal core and circling electrons, featured the requirement for another hypothetical structure.

II. The Dark Body Radiation Issue: Max Planck and Quantization

One of the essential minutes in the introduction of quantum physical science was the goal of the dark body radiation issue, a problem that traditional physical science couldn't satisfactorily make sense of:

1. **The Old style Disappointment:**
   Old style physical science anticipated that the energy transmitted by a dark body ought to increment without bound as the recurrence of radiation expanded (the "bright disaster").
   This error among hypothesis and trial represented a significant test.

2. **Max Planck's Quantum Theory (1900):**

Max Planck presented the progressive thought that energy is quantized, meaning it can exist in discrete parcels or quanta.

Planck's quantization speculation effectively made sense of the dark body radiation bend by matching the noticed information, making it a foundation of quantum hypothesis.

III. The Introduction of Quantum Mechanics: Wave-Molecule Duality

The introduction of quantum material science advanced rapidly with the detailing of wave-molecule duality, an idea that in a general sense changed how we might interpret the idea of particles:

1. **The Double Idea of Light:**
   Tests like the twofold cut try and the photoelectric impact proposed that light shown both wave-like and molecule like properties.
   Albert Einstein's clarification of the photoelectric impact in 1905, which recommended that light comprises of discrete bundles of energy called photons, further underscored the wave-molecule duality.
2. **De Broglie's Wave-Molecule Duality (1924):**

Louis de Broglie expanded the wave-molecule duality idea to issue particles, recommending that particles like electrons additionally display wave-like properties.

This thought established the groundwork for the improvement of quantum mechanics.

IV. The Bohr Model and the Quantum Unrest

The Bohr model of the particle, presented by Niels Bohr in 1913, was a notable advancement that connected traditional and quantum material science:

1. **The Constraints of Old style Nuclear Models:**
   Traditional material science couldn't make sense of the soundness of particles, which ought to transmit energy consistently and breakdown into the core as indicated by old style electromagnetic hypothesis.
2. **The Bohr Model:**

Niels Bohr's model hypothesized that electrons involve explicit, quantized energy levels or circles inside a particle.

This quantization of electron energy levels effectively made sense of the unearthly lines of hydrogen and denoted a critical stage toward grasping nuclear way of behaving.

V. Heisenberg's Vulnerability Standard and the Formalization of Quantum Mechanics

Quantum mechanics took its proper shape with the presentation of Werner Heisenberg's vulnerability standard and the improvement of numerical structures:

1. **Heisenberg's Vulnerability Guideline (1925):**
   Werner Heisenberg's guideline expresses that there are intrinsic cutoff points to all the while knowing the exact position and force of a molecule.
   This rule tested old style determinism and featured the key job of vulnerability in quantum mechanics.

2. **The Schrödinger Condition (1926):**

   Erwin Schrödinger formed the Schrödinger condition, a numerical portrayal of how quantum wavefunctions develop after some time.
   The condition gave a far reaching structure to portraying the way of behaving of particles in quantum frameworks.

VI. The Copenhagen Translation and Quantum Mechanics' Philosophical Ramifications

1. **The Copenhagen Translation:**
   Niels Bohr and Werner Heisenberg fostered the Copenhagen translation of quantum mechanics, which embraced the probabilistic and indeterministic nature of quantum peculiarities.
   This translation declared that the demonstration of estimation assumed an essential part in characterizing the result of quantum occasions.

2. **Einstein's Doubt:**

Albert Einstein still had a few lingering doubts of the probabilistic and indeterministic nature of quantum mechanics. He broadly proclaimed, "God doesn't play dice with the universe," communicating his uneasiness with the vulnerabilities intrinsic in quantum hypothesis.

VII. The EPR Mystery and Bohr's Reaction

1. **The EPR Oddity:**
   The EPR oddity was a psychological study that recommended quantum mechanics considered momentary correspondence between snared particles, in any event, when isolated by significant stretches.
   It tested the fulfillment of quantum mechanics and inferred the presence of "stowed away factors."

2. **Bohr's Reaction:**

Niels Bohr answered the EPR conundrum by protecting the probabilistic and indeterministic nature of quantum mechanics. He contended that the conundrum was a result of traditional reasoning and that quantum mechanics offered a total and exact depiction of quantum peculiarities.

VIII. The Significant Effect of Quantum Material science

The significant effect of quantum material science reaches out a long ways past the domain of logical hypothesis. It has changed how we might interpret the actual world and catalyzed innovative headways:

1. **Innovative Applications:**
   Quantum mechanics has prompted the improvement of various innovations, including semiconductors, lasers, and atomic power. Quantum figuring, with its capability to change data handling, addresses a wilderness in mechanical development.

2. **Philosophical Ramifications:**
   Quantum mechanics challenges traditional instincts about the real world, determinism, and the job of perception, invigorating

continuous philosophical discussions about the idea of the universe.

3. **Quantum Entrapment:**

The peculiarity of quantum ensnarement, fundamental to the EPR oddity, keeps on captivating researchers and thinkers, bringing up issues about the interconnectedness of particles and the idea of spacetime.

1.1A look at the development of quantum physics and the key pioneers like Max Planck.

The improvement of quantum physical science addresses quite possibly of the most progressive and significant logical headway in mankind's set of experiences. This part of

physical science, which manages the way of behaving of particles at the littlest scales, has tested our essential comprehension of the universe. Quantum material science has prompted the advancement of endless innovations and has prepared for leap forwards in different fields, from registering to cryptography. In this investigation, we will dig into the authentic improvement of quantum material science, zeroing in on key trailblazers like Max Planck and their pivotal commitments to this field.

The Introduction of Quantum Physical science

The introduction of quantum physical science can be followed back to the late nineteenth century when researchers started to wrestle with the constraints of old style physical science in making sense of the way of behaving of issue and energy at the nuclear and subatomic levels. As of now, traditional material science, as expressed by Isaac Newton and others, had effectively made sense of a great many regular peculiarities. Nonetheless, when it came to making sense of the way of behaving of particles at the nuclear and subatomic levels, traditional physical science experienced outlandish issues.

One of the basic issues traditional material science confronted was the way of behaving of blackbody radiation. Blackbodies are speculative items that assimilate all episode electromagnetic radiation without reflecting or sending any. When warmed, they transmit a range of radiation, which, as indicated by old style physical science, ought to have been persistent, yet exploratory information showed an alternate reality. Max Planck, a German physicist, was quick to resolve this issue effectively.

Max Planck's Quantum Theory

In 1900, Max Planck proposed a progressive thought that would make way for the improvement of quantum material science. He hypothesized that the energy produced by a blackbody could happen in discrete, quantized units or "quanta." This thought was an extreme takeoff from traditional physical science, which expected that energy could be discharged or retained consistently. Planck presented a numerical recipe, presently known as Planck's radiation regulation, which precisely depicted the noticed range of blackbody radiation. The consistent he presented, presently known as Planck's steady (h), became one of the fundamental constants of quantum material science.

Planck's quantum speculation denoted the introduction of quantum mechanics, as it suggested that energy was not persistent however came in discrete bundles. This pivotal knowledge established the groundwork for the improvement of quantum hypothesis and tested the old style idea of determinism, where the eventual fate of a framework can be anticipated exactly on the off chance that its underlying circumstances are known.

Albert Einstein and the Photoelectric Impact

One more key trailblazer in the advancement of quantum material science was Albert Einstein. In 1905, that very year he distributed his renowned hypothesis of extraordinary relativity, Einstein directed his concentration toward the photoelectric impact, a peculiarity where electrons are transmitted from a material when it is presented to light. Traditional material science anticipated that the energy of the transmitted electrons ought to rely upon the power of the light, yet not on its recurrence.

Einstein, in any case, suggested that light is quantized, made out of discrete bundles of energy he called "photons." He contended that the photoelectric impact must be made sense of on the off chance that light acted as though it were made out of these discrete parcels. This progressive thought was affirmed through analyses, and Einstein was granted the Nobel Prize in Physical science in 1921 for his work on the photoelectric impact. His commitments further hardened the quantum speculation and laid out the wave-molecule duality of light.

Niels Bohr and the Bohr Model

Niels Bohr, a Danish physicist, made critical commitments to the improvement of quantum physical science with his Bohr model of the iota. In 1913, Bohr proposed a model that effectively made sense of the ghastly lines of hydrogen molecules, an issue that had perplexed researchers for quite a long time. As per his model, electrons in molecules possess quantized energy levels, or circles, and can progress between these levels by retaining or transmitting discrete measures of energy.

Bohr's model presented the idea of quantized precise energy and gave a system to grasping the phantom lines of different components. It likewise denoted the start of quantum mechanics' application to the investigation of the particle, showing that old style physical science was lacking for portraying nuclear scale peculiarities.

Werner Heisenberg and the Vulnerability Guideline

Werner Heisenberg, a German physicist, made one of the most significant commitments to quantum material science with his definition of the vulnerability standard in 1927. This standard expresses that it is in a general sense difficult to know both the position and energy of a molecule with outright accuracy all the while. All in all, the more unequivocally we measure one property (e.g., position), the less definitively we can know the other (e.g., energy).

The vulnerability standard tested the old style thought of determinism and acquainted a major cutoff with our capacity to foresee the way of behaving of particles at the quantum level. Heisenberg's work had extensive ramifications, for how we might interpret quantum mechanics as well as for the way of thinking of science. It underscored the probabilistic idea of quantum peculiarities and prompted a more profound philosophical

discussion about the idea of the real world and the job of perception in quantum physical science.

Erwin Schrödinger and Wave Mechanics

Erwin Schrödinger, an Austrian physicist, made huge commitments to quantum material science with his advancement of wave mechanics in 1926. Schrödinger's wave condition depicts the way of behaving of quantum frameworks, like electrons in iotas, with regards to wavefunctions. These wavefunctions address the likelihood disseminations of tracking down particles in various states.

Schrödinger's wave mechanics and Heisenberg's network mechanics, created around a similar time, were viewed as numerically same and turned into the underpinning of quantum mechanics as far as we might be concerned today. Wave mechanics gave a more instinctive and visual understanding of quantum peculiarities, featuring the wave-molecule duality of particles, where they show both wave-like and molecule like properties.

The Introduction of Quantum Electrodynamics (QED)

During the 1940s and 1950s, Richard Feynman, Julian Schwinger, and Tomonaga Shinichiro freely created quantum electrodynamics (QED), a quantum field hypothesis that portrays the electromagnetic connection between charged particles. QED is viewed as quite possibly of the best and exact hypothesis in material science, with forecasts matching exploratory outcomes to unprecedented accuracy.

Feynman, specifically, acquainted a clever methodology with QED called Feynman outlines, which gave a visual portrayal of the intricate collaborations between particles. This graphical strategy upset the field and made quantum field hypothesis more open to physicists. QED's progress in depicting electromagnetic peculiarities prepared for the improvement of quantum field speculations for other crucial powers, like areas of strength for the frail atomic powers.

The Quantum Upset Proceeds

The improvement of quantum physical science didn't stop with the commitments of Planck, Einstein, Bohr, Heisenberg, Schrödinger, and the pioneers behind QED. All things considered, it has proceeded to advance and extend, prompting the improvement of different subfields and applications.

Quantum Mechanics and Molecule Physical science: The Standard Model

During the twentieth hundred years, molecule physicists broadened the standards of quantum mechanics to figure out the way of behaving of subatomic particles and fostered the Standard Model of molecule physical science. This model effectively

portrays the electromagnetic, powerless, and solid atomic powers and predicts the presence of particles like quarks, leptons, and check bosons.

The Standard Model is a demonstration of the force of quantum physical science in making sense of the essential particles and powers of the universe. It has been tried and affirmed through various analyses directed at molecule gas pedals around the world, including the Huge Hadron Collider (LHC) at CERN.

Quantum Mechanics and Innovation

Quantum mechanics significantly affects innovation and designing. Quantum standards support the activity of semiconductors and semiconductors, which are fundamental parts of current gadgets. Furthermore, the advancement of quantum processing can possibly upset registering by playing out specific computations dramatically quicker than traditional PCs.

Quantum mechanics has prompted the production of advances, for example, attractive reverberation imaging (X-ray), which depends on the standards of quantum material science to make nitty gritty pictures of the human body's inside. Quantum cryptography is one more arising field that use the remarkable properties of quantum particles for secure correspondence.

Quantum Mechanics and Central Inquiries

Notwithstanding its gigantic achievement, quantum material science brings up crucial philosophical issues about the idea of the real world and the job of perception. The popular psychological test known as Schrödinger's feline shows the Catch 22s and difficulties of quantum material science. It features that particles can exist in various states all the while until they are noticed.

The idea of quantum entrapment, where particles can become connected so that the condition of one is subject to the condition of another, in any event, when isolated by enormous distances, challenges how we might interpret causality and territory. These and other quantum puzzles keep on being the subject of serious philosophical and logical discussion.

1.2 Explanation of Planck's quantum hypothesis and its implications.

At the beginning of the twentieth hundred years, the domain of material science was on the cusp of a significant change. Old style physical science had effectively made sense of a great many normal peculiarities, yet it confronted outlandish difficulties when it came to grasping the way of behaving of issue and energy at the nuclear and subatomic levels. Max Planck's momentous quantum speculation, proposed in 1900, denoted the start of another period in material science. This speculation not just tended to the well established issue of blackbody radiation yet additionally presented a crucial change in

how we might interpret energy and matter. In this exposition, we will investigate Planck's quantum speculation, its definition, and the significant ramifications it had on the improvement of quantum physical science.

Planck's Quantum Speculation
The Quantum Issue: Blackbody Radiation
One of the key issues that Planck's quantum speculation looked to tackle was the way of behaving of blackbody radiation. A blackbody is a glorified item that retains all occurrence electromagnetic radiation without reflecting or sending any. When warmed, it emanates a range of

radiation, which, as indicated by traditional material science, ought to have been consistent. Nonetheless, exploratory information at the time showed that the range of blackbody radiation didn't match the forecasts of traditional hypothesis.

The Bright Fiasco: Traditional physical science anticipated that the power of blackbody radiation ought to increment without bound as the recurrence of the discharged radiation expanded. This forecast prompted what was known as the "bright fiasco," as it suggested that the energy emanated from a blackbody at high frequencies would be boundlessly huge, an outcome that went against actual reality.

Planck's Quantum Jump: In his quest for an answer for the blackbody radiation issue, Max Planck made a significant takeoff from traditional material science. He recommended that the energy transmitted by a blackbody was not constant however existed in discrete, quantized units or "quanta." These quanta of energy were relative to the recurrence of the radiation, as portrayed by Planck's radiation regulation:

Ramifications of Planck's Quantum Theory

Energy Quantization: Planck's quantum speculation presented the idea of energy quantization, and that implies that energy isn't nonstop however is quantized into discrete bundles or quanta. This was an unmistakable takeoff from the old style view, where energy could fluctuate constantly.

Wave-Molecule Duality: Planck's work added to the improvement of the wave-molecule duality idea. It uncovered that electromagnetic radiation, like light, could show both wave-like and molecule like way of behaving. These quantized bundles of energy, or quanta, are frequently alluded to as "photons."

Birth of Quantum Mechanics: Planck's quantum speculation established the groundwork for quantum mechanics, another system for figuring out the way of behaving of particles at the nuclear and subatomic levels. Quantum mechanics would later turn into the foundation of present day physical science.

Breakdown of Determinism: Old style material science was based on the standard of determinism, where the future way of behaving of a framework could be exactly anticipated in the event that its underlying circumstances were known. Planck's quantum speculation brought a component of indeterminism into the actual world. At the quantum level, certain properties of particles must be portrayed probabilistically, testing the traditional thought of a perfect timing universe.

Rise of Quantum Field Hypothesis: Planck's work made ready for the improvement of quantum field hypothesis, a structure that depicts the crucial powers of nature as fields made out of quantized particles called bosons. Quantum field hypothesis has been critical in making sense of the way of behaving of particles and powers in the quantum domain.

Heisenberg's Vulnerability Guideline: Planck's quantum speculation set up for Werner Heisenberg's detailing of the vulnerability rule. Heisenberg's rule declares that it is on a very basic level difficult to know both the position and force of a molecule with outright accuracy at the same time. This vulnerability is an immediate result of the wave-molecule duality intrinsic in quantum physical science.

Quantum Unrest: Planck's quantum speculation set off a logical upheaval that tested traditional physical science and extended how we might interpret the universe. It prompted the improvement of quantum mechanics, quantum electrodynamics, and quantum field speculations, which aggregately make sense of the way of behaving of particles and powers at the quantum level.

Mechanical Progressions: The improvement of quantum physical science has prompted various innovative headways. Quantum mechanics underlies the activity of semiconductors and semiconductors, fundamental parts of current gadgets. It has additionally brought about advancements, for example, attractive reverberation imaging (X-ray) and quantum registering, promising progressive leap forwards in different fields.

1.3 Early contributions of Einstein and Bohr to quantum theory.

The improvement of quantum hypothesis addresses a urgent crossroads throughout the entire existence of material science, testing old style thoughts of determinism and upsetting comprehension we might interpret the way of behaving of particles at the nuclear and subatomic scales. Albert Einstein and Niels Bohr, two of the most unmistakable researchers of the twentieth hundred years, made early and central commitments to the arising field of quantum mechanics. In this article, we will

investigate the early commitments of Einstein and Bohr to quantum hypothesis, zeroing in on their work on the photoelectric impact, the Bohr model of the particle, and their particular ways to deal with the understanding of quantum peculiarities.

Einstein's Commitments to the Photoelectric Impact

Albert Einstein's work on the photoelectric impact, distributed in 1905, was a fundamental commitment to the improvement of quantum hypothesis. The photoelectric impact is a peculiarity where electrons are radiated from a material when it is presented to light. Old style physical science anticipated that the energy of the radiated electrons ought to rely upon the force of the light, however not on its recurrence.

Einstein's Quantum Speculation: Einstein presented the progressive thought that light is quantized, made out of discrete bundles of energy he called "photons." As indicated by his theory, the energy of a solitary photon is straightforwardly corresponding to its recurrence, as portrayed by the situation

Clarification of the Photoelectric Impact: Einstein's quantum speculation gave a characteristic clarification to the photoelectric impact. He contended that when photons strike a material, they move their energy to electrons inside the material. On the off chance that the energy of a solitary photon is more prominent than the energy expected to eliminate an electron from the material's surface (the work capability), the electron is launched out. This catapulted electron is known as a photoelectron.

Key Ramifications of Einstein's Work: Einstein's clarification of the photoelectric impact had a few significant ramifications:

Energy Quantization: Einstein's work supported Max Planck's concept of energy quantization, showing that energy isn't consistent yet exists in discrete parcels or quanta. This idea tested traditional material science, where energy was accepted to be nonstop.

Molecule Wave Duality: Einstein's speculation that light comprises of quantized particles, photons, added to the arising comprehension of wave-molecule duality in quantum physical science. Photons showed both molecule like conduct in the photoelectric impact and wave-like conduct in peculiarities like diffraction and obstruction.

Affirmation through Investigation: Einstein's expectations with respect to the photoelectric impact were affirmed through tests, areas of strength for offering help for quantized energy levels and the presence of photons. This exploratory approval further set the idea of quantization in quantum hypothesis.

Bohr's Model of the Iota

Niels Bohr, a Danish physicist, made huge early commitments to quantum hypothesis through his improvement of the Bohr model of the particle. In 1913, Bohr proposed a model that effectively made sense of the unearthly lines of hydrogen particles, an issue that had evaded researchers for quite a long time.

Bohr's Proposes: The Bohr model depended on a few key hypothesizes:

Electrons Possess Quantized Energy Levels: Bohr proposed that electrons in molecules involve quantized energy levels, or circles, and can change between these levels by retaining or radiating discrete measures of energy. These energy levels are portrayed by unambiguous energy values.

Discharge and Ingestion of Photons: When an electron changes from a higher energy level to a lower one, it emanates a photon with energy equivalent to the energy distinction between the levels. On the other hand, when an electron retains a photon of the right energy, it can move to a higher energy level.

Rakish Energy Quantization: Bohr presented the possibility that the precise force of electrons in these circles is quantized, meaning it can have explicit, quantized values.

Clarification of Ghastly Lines: Bohr's model effectively made sense of the phantom lines saw in hydrogen particles. At the point when electrons change between energy levels, they emanate or ingest photons with energies relating to the distinctions between the levels. This outcomes in discrete ghostly lines in the electromagnetic range.

Meaning of Bohr's Model: Bohr's model was critical because of multiple factors:

Quantization of Nuclear Energy: It acquainted the idea of quantization with nuclear energy levels, giving a system to grasping the steadiness and conduct of iotas. This was a takeoff from old style physical science, which couldn't make sense of the noticed ghastly lines.

Prescient Power: Bohr's model precisely anticipated the ghastly lines of hydrogen, giving a reasonable and natural clarification for the examples saw in nuclear spectra.

Progress to Quantum Mechanics: While Bohr's model had its restrictions, it established the groundwork for the advancement of more extensive quantum mechanics. Bohr's quantization of rakish force and energy levels foreshadowed the standards of quantum mechanics.

Various Ways to deal with Quantum Understanding

Einstein and Bohr had different philosophical positions on the understanding of quantum peculiarities, prompting critical discussions in the beginning of quantum hypothesis. These discussions keep on affecting conversations about the idea of the quantum world.

Einstein's Authenticity: Einstein held a pragmatist perspective on quantum material science, having faith in the presence of an objective reality free of perception. He broadly expressed, "God doesn't play dice with the universe," communicating his distress with the probabilistic idea of quantum mechanics. Einstein was pained by the indeterminism innate in quantum hypothesis, where the result of a quantum occasion must be anticipated probabilistically.

Bohr's Complementarity: Niels Bohr, then again, fostered a way of thinking of quantum mechanics known as "complementarity." Bohr contended that the way of behaving of particles at the quantum level relied upon the kind of estimation or perception being performed. He recommended that quantum substances, like electrons, could show both molecule like and wave-like properties, contingent upon the exploratory setting.

Goal of the Discussion: The discussion among Einstein and Bohr in regards to the translation of quantum mechanics endured for a long time. It was just settled, as it were, with the approach of the Copenhagen understanding, which orchestrated components of the two perspectives. As per the Copenhagen translation, quantum frameworks exist in a condition of superposition (different potential states) until they are noticed, so, all in all they breakdown into a solitary unmistakable state.

Chapter 2

Einstein's Skepticism

Albert Einstein, perhaps of the most celebrated researcher ever, is frequently hailed for his noteworthy commitments to the hypothesis of relativity and the improvement of present day physical science. In any case, behind his notorious accomplishments lay a profound and withstanding doubt — a scholarly stance that directed his logical requests and formed his perspective. In this investigation, we dig into Einstein's suspicion, its starting points, appearances, and what it meant for his logical interests and philosophical standpoint.

Starting points of Einstein's Distrust

Youth Interest: Einstein showed an unquenchable interest since early on. He frequently addressed customary way of thinking and authority figures, procuring a standing for being a defiant and free mastermind.

Instructive Disappointment: In school, Einstein's discontent with unbending showing techniques and an emphasis on repetition retention drove him to scrutinize the customary schooling system. This disappointment encouraged a solid wariness towards laid out instructive standards.

Family Impact: Einstein's family, especially his uncle Jakob, assumed a huge part in supporting his wariness. Jakob acquainted him with books on science and theory, empowering youthful Albert to freely investigate thoughts.

Early Readings: Einstein's openness to powerful books, for example, Ernst Mach's "The Study of Mechanics" and Immanuel Kant's philosophical works, left an enduring effect. These readings incited him to examine the groundworks of physical science and question the predominant logical ideal models of his time.

Appearances of Einstein's Doubt

Logical Upset: Einstein's most renowned accomplishment, the hypothesis of relativity, tested old style Newtonian physical science. His suspicion towards the overall outright reality drove him to foster a fundamentally new structure for grasping the universe.

Unique Hypothesis of Relativity: In 1905, Einstein presented the extraordinary hypothesis of relativity, which essentially adjusted our ideas of existence. He scrutinized the supremacy of these thoughts and suggested that the laws of physical science are no different for all spectators, no matter what their relative movement.

General Hypothesis of Relativity: Expanding on his wariness, Einstein fostered the overall hypothesis of relativity in 1915. This hypothesis set that gravity emerges from the shape of spacetime, testing the old style gravitational system laid out by Isaac Newton.

Quantum Mechanics: Einstein's incredulity reached out to quantum mechanics, a field he helped pioneer yet later evaluated energetically. While his work on the photoelectric impact and the quantization of light procured him the Nobel Prize, he still had a few some lingering doubts about the probabilistic idea of quantum hypothesis.

Einstein-Podolsky-Rosen Conundrum: in a joint effort with Boris Podolsky and Nathan Rosen, Einstein planned a psychological study in 1935, known as the EPR mystery. This Catch 22 tested the fulfillment of quantum mechanics, contending that it involved "creepy activity a ways off."

Popular Jest: "God Doesn't Play Dice": Einstein's wariness toward the indeterminism of quantum mechanics is typified in his renowned joke, "God doesn't play dice with the universe." He was awkward with the possibility that the basic idea of reality could be administered by unadulterated possibility.

Political and Social Convictions: Einstein's incredulity stretched out to his political and social convictions. He was a frank pundit of militarism, patriotism, and tyranny. His support for demilitarization and pacifism mirrored his doubt towards the disastrous inclinations of human culture.

Pacifism and Against War Activism: Einstein's resistance to The Second Great War and his contribution in different harmony developments showed his suspicion towards the human limit with respect to levelheaded dynamic in issues of war and struggle.

Social equality and Civil rights: Einstein was a supporter for social liberties and civil rights, particularly with regards to racial segregation. His doubt about cultural standards and biases drove him to revolt against racial foul play and isolation.

Strict Convictions: Einstein's strict perspectives were set apart by distrust towards customary strict creed. While he recognized a feeling of miracle and secret about the universe, he dismissed the thought of an individual God who mediates in human issues.

Spinoza's God: Einstein's pantheistic perspectives were frequently compared to those of the savant Baruch Spinoza. He had faith from an enormous perspective of eternality encapsulated in the laws of nature, which he alluded to as "the Bygone One."

Religion as Human Projection: Einstein viewed customary strict convictions as human projections onto the universe. His suspicion drove him to advocate for a judicious, non-obdurate way to deal with otherworldliness.

Impacts on Logical Technique

Standard of Autonomy: Einstein had faith in the significance of scholarly freedom. He kept a sound doubt toward agreement sees and urged researchers to address winning standards. His own earth shattering work on relativity is a demonstration of this rule.

Exact Check: Einstein underscored the meaning of experimental confirmation in science. His wariness towards simply hypothetical develops drove him to look for trial proof to help or invalidate theories. This approach was obvious in his work on the photoelectric impact, where he gave a hypothetical clarification that was subsequently affirmed through tests.

Effortlessness and Style: Einstein esteemed straightforwardness and tastefulness in logical speculations. His wariness towards superfluous intricacy drove him to look for brief and instinctive clarifications for regular peculiarities. This inclination for straightforwardness is exemplified in the well known condition

Logical Uprightness: Einstein kept a pledge to logical trustworthiness and genuineness. His incredulity reached out to the morals of logical exploration, and he unequivocally denounced any control or mutilation of logical discoveries for political or philosophical purposes.

Pluralism and Liberality: Einstein's wariness didn't keep him from thinking about elective perspectives. He stayed open to groundbreaking thoughts and invited scholarly

variety, perceiving that progress frequently emerges from the interaction of alternate points of view.

Tradition of Einstein's Suspicion

Logical Headways: Einstein's suspicion drove him to address laid out physical science, bringing about the definition of the hypothesis of relativity and his commitments to quantum mechanics. His work has molded the direction of current physical science and remains primary in how we might interpret the universe.

Philosophical Request: Einstein's philosophical distrust about the idea of the real world, determinism, and the job of chance has propelled philosophical conversations on the translation of quantum mechanics and the restrictions of human information.

Moral and Moral Backing: Einstein's incredulity towards war, tyranny, and social unfairness energized his promotion for harmony, social liberties, and social value. His principled stand on these issues keeps on rousing developments for positive change.

Instructive Way of thinking: Einstein's disappointment with customary training impacted his perspectives on teaching method. His accentuation on interest, autonomous reasoning, and basic addressing has impacted current instructive way of thinking and ways to deal with educating.

Public Scholarly: Einstein's status as a public scholarly, famous for his logical accomplishments as well as for his moral and moral standards, has made him an image of the dependable utilization of information and the significance of logical doubt.

2.1 Einstein's growing skepticism toward the emerging quantum theory.

Albert Einstein, eminent for his notable commitments to physical science, strangely became one of the main cynics of quantum hypothesis, in spite of assuming an essential part in its initial turn of events. This doubt developed throughout the years as quantum mechanics advanced, and Einstein's issues with its major standards turned out to be more articulated. In this exposition, we will investigate the explanations for Einstein's developing suspicion toward quantum hypothesis, his renowned discussions with individual physicists, and the getting through tradition of his basic position on this progressive part of physical science.

The Introduction of Quantum Mechanics

Prior to digging into Einstein's suspicion, it is fundamental to comprehend the setting where quantum mechanics arose. In the mid twentieth hundred years, physicists were wrestling with the way of behaving of issue and energy at the nuclear and subatomic levels, a domain where old style material science had neglected to give palatable clarifications. Max Planck's quantum speculation, presented in 1900, denoted the introduction of quantum mechanics by proposing that energy was quantized into discrete parcels, or quanta, testing the consistent energy idea of old style physical science.

Quantum mechanics before long found its establishments in crafted by researchers like Niels Bohr, Werner Heisenberg, Max Conceived, and Erwin Schrödinger. These early trailblazers planned the numerical structure that portrayed the way of behaving of particles on a quantum scale, presenting ideas like wave capabilities, wave-molecule duality, and the vulnerability standard. Quantum mechanics immediately acquired acknowledgment because of its surprising progress in making sense of different peculiarities and foreseeing test results.

Einstein's Commitments to Quantum Hypothesis

Albert Einstein's association in the improvement of quantum hypothesis couldn't possibly be more significant. In 1905, that very year he distributed his hypothesis of unique relativity, Einstein made an earth shattering commitment to quantum physical science by making sense of the photoelectric impact. His work, which presented the idea of photons as discrete bundles of light energy, acquired him the Nobel Prize in Material science in 1921.

Einstein's suspicious position on quantum hypothesis didn't arise right away. As a matter of fact, he assumed a functioning part in propelling quantum mechanics during its earliest stages. Notwithstanding, as quantum hypothesis developed, certain parts of it started to conflict with his profoundly held feelings about the idea of the real world and the job of determinism in science. This prompted his developing distrust and his renowned discussions with other driving physicists of his time.

Einstein's Issues with Quantum Mechanics

Determinism: Einstein was a steadfast devotee to the deterministic perspective, which holds that on the off chance that one knows every one of the underlying states of a framework, its future way of behaving can be anticipated with conviction. Quantum mechanics presented a component of indeterminism, with specific actual properties portrayed just probabilistically. This probabilistic nature of quantum hypothesis tested Einstein's philosophical position.

Wave-Molecule Duality: While Einstein was instrumental in the improvement of the idea of wave-molecule duality, he found it thoughtfully disrupting. The possibility that particles like electrons could display both wave-like and molecule like ways of behaving all the while went against his instinctive comprehension of actual reality.

EPR Oddity: In 1935, Einstein, alongside Boris Podolsky and Nathan Rosen, planned what is currently known as the Einstein-Podolsky-Rosen (EPR) Catch 22. This psychological study tested the culmination of quantum mechanics, contending that it suggested immediate "creepy activity a ways off" between snared particles, an idea Einstein saw as profoundly risky.

God Doesn't Play Dice: Einstein broadly communicated his inconvenience with the probabilistic idea of quantum hypothesis by saying, "God doesn't play dice with the universe." He accepted that there should be fundamental deterministic regulations administering quantum peculiarities, regardless of whether they were not yet known.

Discussions and Exchanges

Einstein's developing incredulity toward quantum mechanics prompted energetic discussions and exchanges with his counterparts, most strikingly Niels Bohr. Their trades, frequently held at the Solvay Meetings during the 1920s and 1930s, became amazing throughout the entire existence of material science.

The Bohr-Einstein Discussions: The Bohr-Einstein discusses were described by differentiating perspectives. Niels Bohr, a critical figure in the improvement of quantum mechanics, guarded its probabilistic nature and contended that quantum hypothesis gave a total depiction of the real world, regardless of whether it tested old style determinism.

Complementarity: Bohr presented the idea of complementarity, recommending that different exploratory arrangements could uncover various parts of quantum peculiarities. He contended that wave and molecule portrayals were correlative and that it was impractical to notice the two perspectives in a solitary examination all the while.

Einstein's Psychological tests: Einstein answered with different psychological tests intended to feature what he saw as the reasonable issues of quantum mechanics. The EPR Catch 22, for instance, expected to show that quantum hypothesis was inadequate and proposed the presence of "stowed away factors" that would reestablish determinism.

Heisenberg's Vulnerability Standard: Werner Heisenberg, one more key figure in quantum mechanics, added to the discussions by planning the vulnerability guideline. This standard sets that it is difficult to at the same time know both the position and energy of a molecule with outright accuracy, an idea that Einstein tested.

Effect and Tradition of Einstein's Distrust

Secret Factors: Despite the fact that Einstein's journey for buried factors to reestablish determinism eventually fizzled, his suspicion prodded further examination and the improvement of elective understandings of quantum mechanics, like the de Broglie-Bohm hypothesis.

Ringer's Hypothesis: During the 1960s, physicist John Chime formed Ringer's hypothesis, which showed that any hypothesis in view of stowed away factors, as Einstein had expected, would prompt specific exploratory expectations that contrasted from those of quantum mechanics. Ensuing investigations affirmed the expectations of quantum mechanics, apparently precluding the presence of stowed away factors.

Quantum Snare: Einstein's discussions with Bohr assumed a critical part in how we might interpret quantum ensnarement, a peculiarity in which particles become corresponded so that the condition of one is subject to the condition of another, in any event, when isolated by huge distances. Snare has been tentatively affirmed and is currently bridled for applications in quantum figuring and cryptography.

Philosophical Importance: Einstein's doubt about the understanding of quantum mechanics brought up significant philosophical issues about the idea of the real world, the job of perception, and the restrictions of human information. These inquiries keep on animating philosophical discussions in the way of thinking of science.

2.2 Detailed exploration of his critique of probabilistic and statistical interpretations.

Albert Einstein's significant commitments to physical science stretch out past his work on relativity. He likewise assumed a critical part in the improvement of quantum mechanics, especially in his study of the probabilistic and factual translations of this progressive part of physical science. Einstein's wariness in regards to the culmination of quantum mechanics and probabilistic nature prompted discussions and conversations keep on impacting the field right up 'til now. In this exposition, we will set out on an itemized investigation of Einstein's study of probabilistic and measurable translations in quantum mechanics.

The Probabilistic Idea of Quantum Mechanics

Quantum mechanics, which arose in the mid twentieth hundred years, tested traditional physical science by acquainting a probabilistic system with depict the way of behaving of particles at the nuclear and subatomic levels. The probabilistic idea of quantum mechanics can be summed up in a few key standards:

Superposition: Quantum particles can exist in different states or positions all the while, known as superposition. For example, an electron can exist in a superposition of various energy levels in an iota until it is noticed.

Wave-Molecule Duality: Particles, for example, electrons and photons, display both wave-like and molecule like way of behaving. This duality implies that the way of behaving of quantum elements relies upon whether they are seen as particles or as waves.

Vulnerability Rule: Figured out by Werner Heisenberg, the vulnerability standard expresses that it is difficult to gauge both the position and energy of a molecule with outright accuracy all the while. There is intrinsic vulnerability in these estimations.

Probabilistic Results: Quantum mechanics predicts the result of estimations with regards to probabilities. For instance, it can give the likelihood of finding an electron in a specific energy state or the probability of a molecule displaying a specific way of behaving.

Einstein's Scrutinize of Quantum Mechanics

Determinism: Einstein was a firm devotee to the deterministic perspective, which states that assuming one knows every one of the underlying states of a framework, its future way of behaving can be anticipated with conviction. Quantum mechanics presented a component of indeterminism, where certain actual properties must be depicted probabilistically. This takeoff from determinism was a disputed matter for Einstein.

"God Doesn't Play Dice": Einstein broadly communicated his wariness about the probabilistic idea of quantum mechanics by saying, "God doesn't play dice with the universe." He accepted that there should be basic deterministic regulations overseeing quantum peculiarities, regardless of whether they were not yet known.

Secret Factors: Trying to accommodate quantum mechanics with determinism, Einstein proposed secret factors. These speculative, undetectable boundaries would decide the results of quantum occasions and reestablish determinism to the hypothesis. Einstein, alongside teammates Boris Podolsky and Nathan Rosen, introduced the

Einstein-Podolsky-Rosen (EPR) oddity in 1935 as a psychological study to challenge the culmination of quantum mechanics and propose the presence of stowed away factors.

The Einstein-Podolsky-Rosen (EPR) Conundrum

The EPR conundrum is one of the most renowned commitments to the continuous discussion about the probabilistic and factual translations of quantum mechanics. In their paper, Einstein, Podolsky, and Rosen contended that quantum mechanics couldn't give a total depiction of actual reality, as it appeared to involve "creepy activity a ways off."

The EPR psychological test includes two ensnared particles — particles that have related properties, in any event, when isolated by enormous distances. As per quantum mechanics, estimating one molecule immediately decides the properties of the other, no matter what the distance between them. Einstein and his teammates found this quick connection profoundly alarming and apparently in conflict with the standards of causality and territory.

The EPR paper tested the fulfillment of quantum mechanics by recommending that there should be "covered up factors" administering the properties of particles. These secret factors, whenever known, would consider a deterministic depiction of quantum peculiarities. Einstein's expectation was that future tests could uncover these secret factors, in this way reestablishing determinism to physical science.

In any case, resulting improvements in quantum mechanics and exploratory tests, most prominently John Chime's hypothesis and Ringer tests directed by physicist Alain Perspective, appeared to affirm the probabilistic and non-nearby nature of quantum ensnarement. Ringer's hypothesis showed that any hypothesis in view of stowed away factors would make forecasts that contrasted from those of quantum mechanics, and examinations upheld the expectations of quantum mechanics, apparently precluding stowed away factors.

Einstein's Scrutinize With regards to Translations

Copenhagen Understanding: The Copenhagen translation, supported by Niels Bohr and Werner Heisenberg, embraces the probabilistic idea of quantum mechanics and places that quantum substances exist in a superposition of states until noticed. This understanding underscores the job of the eyewitness in imploding the wave capability and deciding the result of estimations.

2.3His famous quote, "God does not play dice with the universe."

The statement "God doesn't play dice with the universe" is one of the most notable and frequently cited assertions ascribed to Albert Einstein. It typifies Einstein's distrust and uneasiness with specific parts of quantum mechanics, especially its probabilistic and indeterministic nature. In this paper, we will investigate the specific situation, importance, and ramifications of this axiom, analyzing Einstein's philosophical position and its relationship to his commitments to science and the understanding of quantum mechanics.

The Beginning of the Statement

The statement is frequently summarized as "God doesn't play dice with the universe." It is accepted to have been communicated by Einstein during a confidential discussion, and it has become symbolic of his perspectives on the probabilistic idea of quantum mechanics. While there is no exact record of when and to whom he originally offered this expression, it is habitually refered to in conversations about the philosophical ramifications of quantum hypothesis.

Setting: Quantum Mechanics and Einstein's Distrust

To comprehend the meaning of this statement, we should consider the setting where it was made. Quantum mechanics, which arose in the mid twentieth 100 years, presented a significant change in how we might interpret the actual world, especially at the nuclear and subatomic levels. Key standards of quantum mechanics, like superposition, wave-molecule duality, and the vulnerability rule, tested old style determinism and acquainted a probabilistic structure with portray the way of behaving of particles.

Einstein's distress with quantum mechanics was established in his philosophical obligation to determinism — the possibility that assuming that one knows every one of the underlying states of a framework, its future way of behaving can be anticipated with conviction. Quantum mechanics presented a component of indeterminism, where certain actual properties must be portrayed probabilistically. This takeoff from determinism was a disputed matter for Einstein.

Importance and Translation

Dismissal of Irregularity: Einstein's assertion should be visible as a dismissal of the possibility that the crucial cycles of the universe are represented by haphazardness or possibility. He accepted that there should be fundamental deterministic regulations overseeing actual peculiarities, regardless of whether they were not yet known.

Philosophical Abhorrence for Indeterminism: Einstein's distress with indeterminism and probabilistic results in quantum mechanics is clear in this statement. He was

uncomfortable with the possibility that specific occasions must be anticipated probabilistically, instead of with sureness.

The Job of God: The utilization of "God" in the statement is figurative and doesn't be guaranteed to suggest a strict conviction on Einstein's part. All things being equal, it highlights the possibility of a deterministic, methodical universe, which Einstein accepted ought to be open to human grasping through logical standards.

Taking note of that Einstein's utilization of "God" here is figurative and ought not be deciphered as an explanation of strict belief is significant." Rather, it mirrors his journey for a more profound, deterministic comprehension of the universe.

Einstein's Continuous Discussion with Quantum Mechanics

Einstein's evaluate of quantum mechanics was not restricted to this well known expression; it was a focal subject in his continuous discussion with other conspicuous physicists of his time, especially Niels Bohr. Their discussions, frequently held at the Solvay Meetings during the 1920s and 1930s, rotated around the philosophical and calculated underpinnings of quantum mechanics.

Bohr, a critical figure in the improvement of quantum mechanics, protected its probabilistic and indeterministic nature. He contended that quantum hypothesis gave a total depiction of the real world, regardless of whether it tested old style determinism. Bohr's Copenhagen understanding stressed the job of the eyewitness and the probabilistic idea of quantum peculiarities.

Einstein, then again, remained profoundly incredulous of the Copenhagen understanding and the thought that the way of behaving of particles relied upon the demonstration of perception. He accepted that there should be covered up factors or deterministic components that underlie quantum peculiarities, regardless of whether they were not yet known. This doubt prompted the popular Einstein-Podolsky-Rosen

(EPR) oddity in 1935, which tested the culmination of quantum mechanics and recommended the presence of stowed away factors.

The Tradition of the Statement

Philosophical Discussions: The statement keeps on invigorating philosophical conversations about the idea of determinism, arbitrariness, and the restrictions of human information. It prompts inquiries concerning the job of likelihood in portraying the key cycles of the universe.

Translations of Quantum Mechanics: Einstein's suspicion about the probabilistic and indeterministic nature of quantum mechanics has added to continuous discussions

about the understanding of quantum hypothesis. Elective translations, like the many-universes understanding and secret factors speculations, have arisen because of these discussions.

The Idea of Logical Request: The statement highlights the significance of decisive reasoning and distrust in logical request. It advises us that even the most commended researchers can address winning ideal models and rock the boat.

Logical Heritage: While Einstein's protests didn't modify the center standards of quantum mechanics, they improved the talk encompassing the understanding of quantum peculiarities. His commitments to the field stay fantastic, and his inheritance as a basic mastermind and logical visionary perseveres.

Einstein's Complaint: Einstein found the Copenhagen translation profoundly unacceptable because of its dependence on eyewitness subordinate reality and the shortfall of fundamental deterministic regulations.

Many-Universes Understanding: Proposed by Hugh Everett III during the 1950s, the many-universes translation recommends that each quantum occasion brings about a fanning of the universe into numerous equal real factors, each addressing an alternate result of the occasion. This understanding keeps up with determinism yet at the expense of presenting a colossal number of equal universes.

Einstein's Complaint: While the many-universes understanding tends to determinism, Einstein would almost certainly have had a problem with the multiplication of equal universes as a superfluous and lavish speculation.

Secret Factors Speculations: A few physicists, including David Bohm and Louis de Broglie, investigated secret factors hypotheses, which tried to acquaint inconspicuous boundaries with decide the results of quantum occasions. These speculations planned to reestablish determinism to quantum mechanics.

Einstein's Help: Einstein's own proposition of stowed away factors lined up with these hypotheses, as he accepted they held the possibility to accommodate quantum mechanics with his deterministic perspective.

Objective Breakdown Models: Objective breakdown models, like the Ghirardi-Rimini-Weber (GRW) hypothesis, present unconstrained and objective falls of the wave capability as a method for clarifying the change from superposition for a positive state. These models mean to give a deterministic record of quantum estimation.

Einstein's Speculative Help: Albeit these models were created after Einstein's time, they share the objective of bringing deterministic components into quantum mechanics, which might have resounded with his perspectives.

Einstein's Inheritance and the Continuous Discussion

Chime's Hypothesis and Examinations: John Ringer's hypothesis and ensuing tests, for example, those directed by Alain Viewpoint, have generally precluded the chance of stowed away factors that would reestablish determinism to quantum mechanics. These tests affirmed the probabilistic and non-nearby nature of quantum ensnarement.

Philosophical Ramifications: Einstein's scrutinize brought up major issues about the idea of the real world, the job of perception, and the restrictions of human information. These inquiries keep on animating philosophical discussions in the way of thinking of science and the groundworks of quantum mechanics.

Elective Translations: Einstein's suspicion added to the investigation of elective understandings of quantum mechanics, including those that keep a deterministic system. While these translations stay disagreeable, they mirror the persevering through effect of Einstein's scrutinize on the field.

Einstein-Podolsky-Rosen (EPR) Catch 22: The EPR mystery stays a subject of revenue and examination in quantum material science, provoking conversations about the idea of entrapment and non-territory.

3

Chapter 3

Bohr's Copenhagen Interpretation

The Copenhagen translation of quantum mechanics, formed essentially by the Danish physicist Niels Bohr in the mid twentieth 100 years, remains as one of the most compelling and persevering through structures for grasping the quantum world. This understanding has been vital to our perception of the abnormal and irrational peculiarities that oversee the way of behaving of particles on the nuclear and subatomic scales. In this broad investigation, we will dig into Bohr's Copenhagen translation, analyzing its key standards, authentic setting, philosophical ramifications, and its continuous effect on the field of quantum mechanics.

1. **The Quantum Transformation and the Requirement for Translation**

   To see the value in the meaning of the Copenhagen translation, it is fundamental to comprehend the authentic scenery against which it arose. At the turn of the twentieth hundred years, traditional physical science ruled, giving rich clarifications to perceptible peculiarities. Nonetheless, this traditional structure wavered when stood up to with the way of behaving of particles at the quantum level. A progression of exploratory perceptions and revelations, like the photoelectric impact, blackbody radiation, and the quantization of energy by Max Planck, challenged traditional material science and required another hypothetical structure.

Quantum mechanics, a weighty hypothesis that arose in the mid twentieth hundred years, tried to make sense of the way of behaving of particles at the nuclear and subatomic scales. Key advancements included:

Planck's Quantum Speculation: In 1900, Max Planck presented the possibility that energy is quantized into discrete bundles, or "quanta." This speculation established the groundwork for quantum mechanics by testing the traditional thought of constant energy.

Wave-Molecule Duality: The wave-molecule duality guideline, created by Louis de Broglie, Werner Heisenberg, and Erwin Schrödinger, proposed that particles like electrons show both wave-like and molecule like way of behaving. This duality essentially adjusted how we might interpret matter.

Heisenberg's Vulnerability Guideline: Werner Heisenberg's vulnerability rule, planned in 1927, laid out a crucial breaking point to the accuracy with which certain sets of

properties, like position and force, could be all the while known. This rule brought inborn vulnerability into quantum estimations.

Quantization of Circles: Niels Bohr's model of the hydrogen iota, created in 1913, made sense of the quantization of electron circles and effectively anticipated the phantom lines of hydrogen. This model addressed a scaffold among traditional and quantum physical science.

As quantum mechanics progressed, obviously the hypothesis introduced an on a very basic level different perspective on the real world, testing old style determinism and presenting probabilistic results. This prompted the requirement for translations

— philosophical structures that could figure out the numerical formalism of quantum mechanics and give a reasonable comprehension of quantum peculiarities.

2. **Verifiable Setting and the Copenhagen Translation's Starting points**

The Copenhagen understanding, which arose fundamentally from the cooperative endeavors of Niels Bohr and Werner Heisenberg, tracked down its underlying foundations in a progression of conversations and discussions among physicists during the 1920s. The expression "Copenhagen translation" was authored on the grounds that a significant number of these conversations occurred at the Niels Bohr Organization in Copenhagen, Denmark.

Key parts of the authentic setting include:

Comfort of Traditional Determinism: Numerous physicists, including Albert Einstein and Erwin Schrödinger, were awkward with the indeterministic and probabilistic nature of quantum mechanics. They tried to find deterministic clarifications or secret factors that could reestablish old style determinism to the hypothesis.

Bohr's Complementarity Rule: Niels Bohr presented the idea of complementarity, which accentuated that different exploratory arrangements could uncover various parts of quantum peculiarities. This standard recognized that wave and molecule portrayals were reciprocal and that it was impractical to notice the two viewpoints in a solitary trial all the while.

Heisenberg's Framework Mechanics: Werner Heisenberg's network mechanics and the Schrödinger wave condition gave two distinct numerical details of quantum mechanics. These details seemed, by all accounts, to be in conflict, bringing up issues about the idea of quantum reality.

The Vulnerability Rule: Heisenberg's vulnerability standard, which essentially tested old style determinism, provoked banters about the idea of the real world and the job of estimation in quantum mechanics.

3. **Key Standards of the Copenhagen Translation**

Wave-Molecule Duality: The Copenhagen understanding hugs wave-molecule duality, which places that quantum substances, for example, electrons, show both wave-like and molecule like way of behaving. Contingent upon the exploratory setting, particles can appear as waves or particles.

Complementarity: The guideline of complementarity is key to the Copenhagen translation. It recognizes that different exploratory arrangements uncover various parts of quantum peculiarities. For instance, tests might zero in on the wave-like or molecule like properties of particles, however not both at the same time.

Probabilistic Nature: The Copenhagen understanding acknowledges the probabilistic idea of quantum mechanics. That's what it attests, preceding estimation, quantum substances exist in superpositions of potential states, and their way of behaving must be portrayed probabilistically.

Job of the Spectator: In the Copenhagen translation, the job of the eyewitness is foremost. The demonstration of perception implodes the quantum wave capability, deciding the result of an estimation. The eyewitness assumes a basic part in characterizing reality at the quantum level.

No Secret Factors: The Copenhagen understanding oddballs the presence of stowed away factors that could give a total deterministic portrayal of quantum peculiarities. It keeps up with that quantum mechanics gives a total and exact depiction of the real world, regardless of whether it is innately probabilistic.

4. **The Job of Estimation and Wave Capability Breakdown**

Fundamental to the Copenhagen understanding is the idea of estimation and the related breakdown of the quantum wave capability. The wave capability depicts the probabilistic dispersion of potential expresses a quantum substance can involve.

At the point when a quantum estimation is made, the wave capability implodes to a particular state relating to the result of the estimation. This breakdown is innately probabilistic; it can't be anticipated with conviction which result will happen. The demonstration of estimation characterizes the actual truth of the quantum substance.

The Copenhagen translation attests that quantum elements, preceding estimation, exist in superpositions of potential states. For instance, an electron in an unmeasured state might exist in a superposition of numerous positions or energy levels all the while. Just upon estimation does it expect a distinct position or, not entirely settled by the probabilistic idea of the wave capability.

This idea of wave capability breakdown and the job of estimation in characterizing reality has significant philosophical ramifications, including inquiries regarding the idea of perception and the spectator's part in deciding the result of quantum occasions.

5. **Einstein's Complaints and the EPR Mystery**

Prominent among the pundits of the Copenhagen understanding was Albert Einstein, who was awkward with its probabilistic and indeterministic nature. Einstein, alongside colleagues Boris Podolsky and Nathan Rosen, planned what is currently known as the Einstein-Podolsky-Rosen (EPR) Catch 22 of every 1935.

The EPR conundrum intended to challenge the culmination of quantum mechanics and propose the presence of stowed away factors. That's what it set, as per quantum mechanics, caught particles — particles that have related properties — could display momentary "creepy activity a ways off." all in all, an estimation made on one molecule could promptly decide the properties of another, in any event, when isolated by huge distances. This obvious non-territory disturbed Einstein, who accepted that actual **impacts shouldn't engender quicker than the speed of light.**

The EPR conundrum raised doubt about the culmination of quantum mechanics, proposing that there may be hidden factors or "components of the real world" that decide the results of estimations. Einstein's expectation was that future tests could uncover these secret factors, subsequently reestablishing determinism to material science.

In any case, ensuing advancements in quantum mechanics, most eminently John Ringer's hypothesis and Chime tests directed by physicist Alain Viewpoint, appeared to affirm the probabilistic and non-neighborhood nature of quantum ensnarement. Chime's hypothesis showed that any hypothesis in light of stowed away factors would make expectations that contrasted from those of quantum mechanics, and trials upheld the forecasts of quantum mechanics, apparently precluding stowed away factors.

Einstein's complaints, while powerful and interesting, didn't generally adjust the center standards of the Copenhagen understanding or quantum mechanics. The EPR oddity and Ringer's hypothesis remain subjects of revenue and examination in the field of quantum physical science, adding to progressing banters about the idea of trap and non-territory.

6. **Philosophical Ramifications and Translations of the Copenhagen Understanding**

The Job of the Eyewitness: The Copenhagen translation highlights the focal job of the spectator in characterizing reality at the quantum level. This brings up philosophical issues about the connection among cognizance and the actual world, as well as the idea of perception itself.

The Idea of Likelihood: The probabilistic idea of quantum mechanics challenges traditional determinism and brings key vulnerability into actual cycles. This has prompted banters about the idea of likelihood and its part in depicting the way of behaving of quantum substances.

Complementarity and Wave-Molecule Duality: The idea of complementarity features that different trial viewpoints uncover various parts of quantum peculiarities. This has suggestions for how we might interpret how we conceptualize and display actual reality.

Determinism versus Indeterminism: The Copenhagen understanding's acknowledgment of probabilistic results diverges from Einstein's inclination for deterministic clarifications. This philosophical gap reflects more extensive discussions
about determinism and indeterminism in science and reasoning.

The Copenhagen translation additionally exists close by elective understandings of quantum mechanics, each offering its own point of view on the idea of quantum reality. A portion of these translations include:

Many-Universes Translation: Proposed by Hugh Everett III, the many-universes understanding recommends that each quantum occasion brings about a stretching of the universe into various equal real factors, each addressing an alternate result of the occasion. This translation keeps up with determinism yet at the expense of presenting a tremendous number of equal universes.

Secret Factors Hypotheses: A few physicists, including David Bohm and Louis de Broglie, investigated secret factors speculations, which acquaint imperceptible boundaries with decide the results of quantum occasions. These hypotheses mean to reestablish determinism to quantum mechanics.

Objective Breakdown Models: Objective breakdown models, like the Ghirardi-Rimini-Weber (GRW) hypothesis, present unconstrained and objective falls of
the wave capability as a method for clarifying the progress from superposition for a distinct state. These models intend to give a deterministic record of quantum estimation.

Quantum Bayesianism: Quantum Bayesianism, or QBism, centers around the abstract idea of quantum likelihood. It places that quantum mechanics depicts a specialist's very own convictions about actual frameworks, instead of an objective reality.

Every understanding offers a one of a kind point of view on

quantum peculiarities and addresses the philosophical and calculated difficulties presented by quantum mechanics.

7. Influence and Continuous Significance

The Copenhagen understanding of quantum mechanics lastingly affects the field of material science and the more extensive academic local area. Its key standards, including wave-molecule duality, complementarity, and the job of the eyewitness, have become basic to how we might interpret quantum peculiarities.

The effect of the Copenhagen understanding stretches out to a few regions:

Quantum Innovation: The standards of quantum mechanics, as clarified by the Copenhagen translation, have made ready for the improvement of quantum advances, including quantum registering, quantum cryptography, and quantum sensors. These advancements influence the extraordinary properties of quantum frameworks.

Philosophical Talk: The Copenhagen understanding keeps on animating philosophical conversations about the idea of the real world, the job of the spectator, and the ramifications of quantum indeterminacy. It has added to the way of thinking of science and the way of thinking of psyche.

Trial Confirmation: Numerous exploratory tests and perceptions have affirmed the expectations of quantum mechanics, offering observational help for the standards of the Copenhagen translation. These analyses have shown the truth of quantum entrapment and the probabilistic idea of quantum estimations.

Instructive Importance: The Copenhagen translation and its related standards are shown in physical science educational programs around the world. They act as fundamental ideas for understudies concentrating on quantum mechanics and quantum physical science.

3.1 Introduction to Niels Bohr's Copenhagen interpretation of quantum mechanics.

Niels Bohr, a Danish physicist, made huge commitments to the improvement of quantum mechanics in the mid twentieth 100 years. His momentous experiences and thoughts established the groundwork for what is known as the Copenhagen translation of quantum mechanics. This translation plays had a vital impact in how we might interpret the quantum world, tending to the unusual and confusing way of behaving of particles at the nuclear and subatomic scales. In this paper, we will dig into the fundamental ideas and standards of Niels Bohr's Copenhagen translation, investigating its verifiable setting, key precepts, and persevering through influence on the field of physical science.

1. The Development of Quantum Mechanics

The mid twentieth century denoted a time of progressive change in the field of material science. Old style material science, which had effectively depicted the way of behaving of naturally visible articles, experienced unfavorable difficulties when applied to the domain of the tiny — molecules, electrons, and photons. A progression of exploratory perceptions and revelations opposed old style material science, requiring the improvement of another hypothetical structure: quantum mechanics.

Key advancements that prompted the introduction of quantum mechanics include:

Planck's Quantum Speculation: In 1900, Max Planck suggested that energy is quantized, meaning it exists in discrete bundles called "quanta." This thought tested the old style idea of non-stop energy.

Wave-Molecule Duality: Physicists like Louis de Broglie, Werner Heisenberg, and Erwin Schrödinger presented the idea of wave-molecule duality, proposing that particles, for example, electrons, show both wave-like and molecule like way of behaving.

Heisenberg's Vulnerability Guideline: In 1927, Werner Heisenberg figured out the vulnerability rule, which laid out central cutoff points to the accuracy with which certain sets of properties, similar to position and force, could be at the same time known.

Bohr's Model of the Hydrogen Molecule: In 1913, Niels Bohr fostered a model of the hydrogen particle that effectively made sense of the quantization of electron circles and anticipated the ghostly lines of hydrogen. This model addressed a scaffold among old style and quantum physical science.

As quantum mechanics progressed, it became clear that it introduced a profoundly unique perspective contrasted with traditional material science. Quantum mechanics presented probabilistic results, wave-molecule duality, and the thought that particles could exist in superpositions of states until estimated. These original elements represented the requirement for translations — philosophical systems that could get a handle on the numerical formalism of quantum mechanics and give an intelligible comprehension of quantum peculiarities.

2. **Authentic Setting and the Copenhagen Understanding's Beginnings**

The Copenhagen translation of quantum mechanics, frequently just alluded to as the Copenhagen understanding, arose during a progression of conversations and discussions among physicists during the 1920s. These discussions, which occurred at the Niels Bohr Foundation in Copenhagen, Denmark, prompted the detailing of key standards and fundamentals that portray Bohr's understanding.

Key parts of the authentic setting include:

Difficulties to Determinism: Physicists of the time, including Albert Einstein and Erwin Schrödinger, were uncomfortable with the indeterministic and probabilistic nature of quantum mechanics. They tried to find deterministic clarifications or secret factors that could reestablish old style determinism to the hypothesis.

Bohr's Complementarity Guideline: Niels Bohr presented the idea of complementarity, which underlined that different trial arrangements could uncover various parts of quantum peculiarities. This rule recognized that wave and molecule depictions were reciprocal and that it was unrealistic to notice the two perspectives in a solitary examination at the same time.

Heisenberg's Framework Mechanics: Werner Heisenberg's lattice mechanics and Erwin Schrödinger's wave condition gave two particular numerical details of quantum mechanics. These plans had all the earmarks of being in conflict, bringing up issues about the idea of quantum reality.

The Vulnerability Standard: Heisenberg's vulnerability rule, which generally tested traditional determinism, provoked banters about the idea of the real world and the job of estimation in quantum mechanics.

3. **Key Standards of the Copenhagen Understanding**

The Copenhagen understanding of quantum mechanics is described by a few vital standards and principles that address the characteristics of quantum peculiarities:

Wave-Molecule Duality: The Copenhagen translation embraces wave-molecule duality, which places that quantum substances, for example, electrons, show both wave-like and molecule like way of behaving. Contingent upon the exploratory setting, particles can appear as waves or particles.

Complementarity: Fundamental to the Copenhagen translation is the rule of complementarity. This rule recognizes that different trial arrangements uncover various parts of quantum peculiarities. For instance, tests might zero in on the wave-like or molecule like properties of particles, however not both all the while.

Probabilistic Nature: The Copenhagen understanding acknowledges the probabilistic idea of quantum mechanics. That's what it affirms, preceding estimation, quantum substances exist in superpositions of potential states, and their way of behaving must be depicted probabilistically.

Job of the Eyewitness: In the Copenhagen translation, the job of the spectator is principal. The demonstration of perception falls the quantum wave capability, deciding the result of an estimation. The onlooker assumes a key part in characterizing reality at the quantum level.

No Secret Factors: The Copenhagen understanding oddballs the presence of stowed away factors that could give a total deterministic depiction of quantum peculiarities. It keeps up with that quantum mechanics gives a total and precise depiction of the real world, regardless of whether it is innately probabilistic.

These standards, frequently connected with Niels Bohr's work, structure the groundwork of the Copenhagen translation and have significant ramifications for how we might interpret the quantum world.

4. **The Job of Estimation and Wave Capability Breakdown**

    At the center of the Copenhagen understanding falsehoods the idea of estimation and the related breakdown of the quantum wave capability. The wave capability is a numerical build that portrays the probabilistic conveyance of potential expresses a quantum element can possess.

    At the point when a quantum estimation is made, the wave capability falls to a particular state relating to the result of the estimation. This breakdown is innately probabilistic, meaning it can't be anticipated with sureness which result will happen. The demonstration of estimation characterizes the actual truth of the quantum element.

    The Copenhagen translation declares that quantum substances, before estimation, exist in superpositions of potential states. For instance, an electron in an unmeasured state might exist in a superposition of numerous positions or energy levels at the same time. Just upon estimation does it expect an unmistakable position or, not entirely set in stone by the probabilistic idea of the wave capability.

    This idea of wave capability breakdown and the job of estimation in characterizing reality have significant philosophical ramifications, including inquiries regarding the idea of perception, the onlooker's job in deciding the result of quantum occasions, and the connection among awareness and actual reality.

5. **Einstein's Complaints and the EPR Oddity**

Albert Einstein, one of the most unmistakable physicists of the twentieth hundred years, was a remarkable pundit of the Copenhagen understanding. His complaints were established in his distress with the probabilistic and indeterministic nature of quantum mechanics. In a joint effort with Boris Podolsky and Nathan Rosen, Einstein figured out what is currently known as the Einstein-Podolsky-Rosen (EPR) mystery in 1935.

The EPR mystery planned to challenge the culmination of quantum mechanics and recommend the presence of stowed away factors. That's what it set, as per quantum mechanics, snared particles — particles that have related properties — could show quick "creepy activity a ways off." at the end of the day, an estimation made on one molecule could momentarily decide the properties of another, in any event, when isolated by immense distances. This obvious non-area disturbed Einstein, who accepted that actual impacts shouldn't engender quicker than the speed of light.

Notwithstanding, ensuing advancements in quantum mechanics, most eminently John Ringer's hypothesis and Chime tests directed by physicist Alain Angle, appeared to affirm the probabilistic and non-nearby nature of quantum snare. Chime's hypothesis exhibited that any hypothesis in light of stowed away factors would make expectations that contrasted from those of quantum mechanics, and trials upheld the forecasts of quantum mechanics, apparently precluding stowed away factors.

Einstein's complaints, while persuasive and intriguing, didn't in a general sense modify the center standards of the Copenhagen understanding or quantum mechanics. The EPR mystery and

Ringer's hypothesis remain subjects of revenue and examination in the field of quantum physical science, adding to progressing banters about the idea of snare and non-territory.

6. **Philosophical Ramifications and Understandings of the Copenhagen Translation**

The Job of the Spectator: The Copenhagen understanding highlights the focal job of the onlooker in characterizing reality at the quantum level. This brings up philosophical issues about the connection among cognizance and the actual world, as well as the idea of perception itself.

The Idea of Likelihood: The probabilistic idea of quantum mechanics challenges old style determinism and brings principal vulnerability into actual cycles. This has prompted banters about the idea of likelihood and its job in portraying the way of behaving of quantum substances.

Complementarity and Wave-Molecule Duality: The idea of complementarity features that different exploratory points of view uncover various parts of quantum peculiarities. This has suggestions for how we might interpret how we conceptualize and display actual reality.

Determinism versus Indeterminism: The Copenhagen understanding's acknowledgment of probabilistic results appears differently in relation to Einstein's inclination for deterministic clarifications. This philosophical gap reflects more extensive discussions about determinism and indeterminism in science and reasoning.

The Copenhagen translation additionally exists close by elective understandings of quantum mechanics, each offering its own point of view on the idea of quantum reality. A portion of these understandings include:

Many-Universes Understanding: Proposed by Hugh Everett III, the many-universes translation recommends that each quantum occasion brings about a spreading of the universe into various equal real factors, each addressing an alternate result of the occasion. This translation keeps up with determinism however at the expense of presenting a gigantic number of equal universes.

Secret Factors Hypotheses: A few physicists, including David Bohm and Louis de Broglie, investigated secret factors speculations, which acquaint undetectable boundaries with decide the results of quantum occasions. These speculations intend to reestablish determinism to quantum mechanics.

Objective Breakdown Models: Objective breakdown models, like the Ghirardi-Rimini-Weber (GRW) hypothesis, present unconstrained and objective implodes of the wave capability as a method for making sense of the change from
superposition for a positive state. These models plan to give a deterministic record of quantum estimation.

Quantum Bayesianism: Quantum Bayesianism, or QBism, centers around the abstract idea of quantum likelihood. It places that quantum mechanics depicts a specialist's very own convictions about actual frameworks, as opposed to an objective reality.

Every understanding offers an extraordinary viewpoint on quantum peculiarities and addresses the philosophical and reasonable difficulties presented by quantum mechanics.

7. Influence and Progressing Pertinence

The Copenhagen understanding of quantum mechanics lastingly affects the field of material science and the more extensive academic local area. Its key standards, including wave-molecule duality, complementarity, and the job of the eyewitness, have become principal to how we might interpret quantum peculiarities.

The effect of the Copenhagen understanding reaches out to a few regions:

Quantum Innovation: The standards of quantum mechanics, as clarified by the Copenhagen translation, have made ready for the advancement of quantum innovations, including quantum processing, quantum cryptography, and quantum sensors. These advancements influence the interesting properties of quantum frameworks.

Philosophical Talk: The Copenhagen understanding keeps on animating philosophical conversations about the idea of the real world, the job of the eyewitness, and the ramifications of quantum indeterminacy. It has added to the way of thinking of science and the way of thinking of psyche.

Exploratory Confirmation: Numerous trial tests and perceptions have affirmed the expectations of quantum mechanics, offering experimental help for the standards of the Copenhagen translation. These examinations have shown the truth of quantum ensnarement and the probabilistic idea of quantum estimations.

Instructive Importance: The Copenhagen understanding and its related standards are shown in material science educational plans around the world. They act as fundamental ideas for understudies concentrating on quantum mechanics and quantum physical science.

3.2 Explanation of the principles and key ideas behind his interpretation.

Niels Bohr's Copenhagen understanding of quantum mechanics is an essential system that looks to get a handle on the particular and strange way of behaving of particles on the nuclear and subatomic scales. It presents a few critical standards and thoughts that shape how we might interpret the quantum world. In this exposition, we will give a top to bottom clarification of these standards and key thoughts, revealing insight into the center precepts of the Copenhagen translation.

1. **Wave-Molecule Duality:** One of the focal standards of the Copenhagen understanding is the idea of wave-molecule duality. This thought, which had been created by before physicists like Louis de Broglie, Werner Heisenberg, and Erwin Schrödinger, places that quantum substances, for example, electrons, display both wave-like and molecule like way of behaving.

   Clarification: In old style physical science, particles were considered as unmistakable, limited elements with clear properties. In any case, quantum mechanics tested this thought by recommending that particles could likewise act as waves. This duality implies that electrons, for instance, could show wave-like obstruction designs while going through a twofold cut explore. However, they could likewise appear as particles with unequivocal positions when estimated. The Copenhagen understanding hugs this duality and stresses that it is a principal part of quantum reality.

Suggestion: Wave-molecule duality challenges our old style instincts about the idea of particles. It recommends that the way of behaving of quantum elements is intrinsically probabilistic and relies upon the setting of the estimation.

2. **Complementarity:** The standard of complementarity is one more key idea in Bohr's translation. Complementarity affirms that different exploratory arrangements can uncover various parts of quantum peculiarities, and these viewpoints are corresponding in nature.

Clarification: Complementarity emerges from the possibility that wave and molecule portrayals are different sides of a similar quantum coin. In specific examinations, we might decide to zero in on the wave-like properties of particles, while in others, we might stress their molecule like way of behaving. Be that as it may, it is preposterous to notice the two viewpoints in a solitary trial at the same time. Complementarity highlights that different trial viewpoints give unique, yet similarly substantial, bits of knowledge into quantum peculiarities.

Suggestion: Complementarity challenges the idea of a solitary, objective reality. It recommends that the truth of quantum elements is setting reliant and that different trial perspectives uncover various aspects of that reality. This idea has significant ramifications for how we might interpret the idea of actual reality.

3. **Probabilistic Nature:** The Copenhagen understanding recognizes the probabilistic idea of quantum mechanics. That's what it attests, preceding estimation, quantum elements exist in superpositions of potential states, and their way of behaving must be depicted probabilistically.

   Clarification: Quantum superposition alludes to the state wherein a quantum substance, similar to an electron, can exist in a straight mix of numerous potential states at the same time. For example, an electron might be in a superposition of different positions or energy levels. The demonstration of estimation implodes this superposition to a particular state with a specific likelihood. The probabilistic idea of quantum mechanics implies that it is difficult to foresee with sureness the result of an estimation; all things being equal, we can portray the probabilities of different results.

   Suggestion: Quantum indeterminacy challenges traditional determinism and brings basic vulnerability into actual cycles. That's what it suggests, at the quantum level, nature is innately eccentric, and deterministic expectations are supplanted by probabilistic ones.

4. **Job of the Onlooker:** In the Copenhagen understanding, the job of the eyewitness is vital. The demonstration of perception implodes the quantum wave capability, deciding the result of an estimation. The spectator assumes a principal part in characterizing reality at the quantum level.

   Clarification: As per Bohr's understanding, the demonstration of estimation is certainly not a uninvolved cycle; it effectively takes part in characterizing the result. Before estimation, quantum substances exist in a superposition of potential expresses, each with related probabilities. At the point when a perception is made, the wave capability breakdowns, and the quantum element expects a clear state comparing to the estimation result. This suggests that the onlooker's decision of what to quantify and the demonstration of estimation itself straightforwardly impact the truth of the quantum framework.

Suggestion: The job of the eyewitness brings up significant philosophical issues about the idea of cognizance, the connection between the spectator and the noticed, and the subjectivity innate in quantum estimations. It recommends that human perception is an essential piece of the quantum reality.

5. **No Secret Factors:** The Copenhagen understanding oddballs the presence of stowed away factors that could give a total deterministic depiction of quantum peculiarities. It keeps up with that quantum mechanics gives a total and exact depiction of the real world, regardless of whether it is innately probabilistic.

   Clarification: Stowed away factors are speculative, imperceptible properties or boundaries that could, in principle, decide the results of quantum estimations in a deterministic way. In any case, the Copenhagen translation affirms that there are no such secret factors. All things being equal, quantum mechanics offers a total and independent depiction of quantum peculiarities. It places that the probabilistic idea of quantum estimations isn't because of our obliviousness of stowed away factors yet is an inborn element of the quantum world.

   Suggestion: Dismissing stowed away factors builds up the possibility that quantum mechanics gives a total and self-steady portrayal of the real world. It proposes that the intrinsic probabilistic nature of quantum peculiarities isn't a limit of our insight yet a fundamental component of the quantum universe.

6. **Wave Capability Breakdown:** Key to the Copenhagen understanding is the idea of wave capability breakdown. The wave capability portrays the probabilistic dispersion of potential expresses a quantum element can possess. At the point when an estimation is made, the wave capability falls to a particular state.

Clarification: The wave capability is a numerical develop that encodes every one of the potential expresses a quantum substance can be in, alongside their related probabilities. At the point when an estimation is played out, the wave capability breakdowns, and the quantum substance expects a particular state comparing to the estimation result. This breakdown is intrinsically probabilistic; it can't be anticipated with sureness which result will happen.

Suggestion: Wave capability breakdown is a major part of quantum mechanics that features the job of estimation in characterizing reality. It highlights that quantum elements exist in a condition of probability until noticed, and the demonstration of perception takes shape their state.

In outline, Niels Bohr's Copenhagen translation of quantum mechanics presents a few major standards and key thoughts that support how we might interpret the quantum world. These standards incorporate wave-molecule duality, complementarity, the probabilistic idea of quantum mechanics, the job of the onlooker, the dismissal of stowed away factors, and the idea of wave capability breakdown. Together, these thoughts give a structure to grasping the odd and confusing way of behaving of particles at the quantum level, offering experiences into the idea of reality itself. Bohr's translation lastingly affects the field of physical science and keeps on forming our investigation of the quantum universe.

3.3 The reception and controversies surrounding Bohr's interpretation.

Niels Bohr's Copenhagen translation of quantum mechanics, with its accentuation on complementarity, wave-molecule duality, and the focal job of the spectator, altogether molded how we might interpret the quantum world. Nonetheless, it was not without its portion of contentions and discussions inside established researchers. This exposition investigates the gathering and discussions encompassing Bohr's understanding, revealing insight into the different viewpoints that arose in light of this compelling system.

1. **Early Gathering and Acknowledgment:**

   At the point when Niels Bohr originally introduced his Copenhagen understanding during the 1920s, it met with a blended however commonly open reaction from established researchers. A few physicists viewed the translation's standards as rationally fulfilling, as they gave an intelligible system to figuring out the cryptic way of behaving of quantum substances.

   Bohr's thoughts offered a method for accommodating the wave-molecule duality and the probabilistic idea of quantum mechanics. His model of the hydrogen molecule, which effectively made sense of the quantization of electron circles and the ghastly lines of hydrogen, loaned exact help to his understanding.

2. **Einstein's Scrutinize and the EPR Oddity:**

   One of the most eminent pundits of the Copenhagen understanding was Albert Einstein. Einstein, alongside Boris Podolsky and Nathan Rosen, figured out what is currently known as the Einstein-Podolsky-Rosen (EPR) Catch 22 of every 1935. The EPR conundrum planned to challenge the fulfillment and philosophical ramifications of quantum mechanics.

   Einstein protested the probabilistic and indeterministic nature of quantum mechanics, broadly expressing, "God doesn't play dice with the universe." He accepted that there should be basic "stowed away factors" that could give a deterministic clarification to quantum peculiarities.

   The EPR oddity fixated on the idea of snare, in which two particles become corresponded so that estimations on one molecule promptly decide the properties of the other, in any event, when isolated by immense distances. This evident "creepy activity a ways off" tangled with Einstein's view that no data or impact could travel quicker than the speed of light.

   While the EPR conundrum didn't straightforwardly challenge all parts of the Copenhagen understanding, it brought up issues about the culmination of quantum mechanics and the philosophical ramifications of quantum snare.

3. **The Bohr-Einstein Discussions:**

The conflicts between Niels Bohr and Albert Einstein with respect to the translation of quantum mechanics prompted a progression of vivacious discussions. These discussions occurred during the popular Solvay Meetings during the 1920s and 1930s, where driving physicists accumulated to talk about central inquiries in physical science.

Bohr guarded the Copenhagen translation, underlining that quantum mechanics gave a total and self-reliable depiction of the real world. That's what he contended, while the probabilistic idea of quantum mechanics may be disrupting, it was a principal part of the quantum world.

Einstein, then again, stayed unfaltering in his faith in a deterministic clarification. He battled that quantum mechanics was deficient and that there should exist stowed away factors that could represent the probabilistic results saw in quantum tests.

The Bohr-Einstein discusses were instrumental in featuring the profound philosophical partitions inside the material science local area with respect to the translation of quantum mechanics. While Bohr's understanding won as far as inescapable acknowledgment, Einstein's protests kept on impacting conversations about the groundworks of quantum physical science.

4. **Improvements in Chime's Hypothesis and Angle's Examinations:**

During the 1960s and 1970s, physicist John Chime figured out a hypothesis that introduced another point of view on the discussion among Bohr and Einstein. Ringer's hypothesis showed that any hypothesis in light of stowed away factors would make expectations that varied from those of quantum mechanics.

Ensuing tests directed by Alain Angle during the 1980s appeared to affirm the expectations of quantum mechanics, suggesting that the relationships saw between entrapped particles couldn't be made sense of by stowed away factors. This offered observational help for the probabilistic and non-neighborhood parts of quantum entrapment, lining up with Bohr's translation. While these improvements didn't determine the philosophical discussions encompassing the translation of quantum mechanics, they tested the practicality of stowed away factors hypotheses and supported the standards of the Copenhagen understanding.

5. **Elective Translations of Quantum Mechanics:**

The contentions and discussions encompassing the Copenhagen translation led to various elective understandings of quantum mechanics. A portion of these understandings looked to resolve the issues raised by Einstein and offered different philosophical points of view on quantum reality. These options include:

Many-Universes Translation: Proposed by Hugh Everett III, the many-universes understanding recommends that each quantum occasion brings about a stretching of the universe into various equal real factors, each addressing an alternate result of the occasion. This translation keeps up with determinism however presents a huge number of equal universes.

Secret Factors Speculations: Physicists like David Bohm and Louis de Broglie investigated secret factors hypotheses, which acquaint inconspicuous boundaries with decide the results of quantum occasions. These hypotheses intend to reestablish determinism to quantum mechanics.

Objective Breakdown Models: Objective breakdown models, like the Ghirardi-Rimini-Weber (GRW) hypothesis, present unconstrained and objective falls of the wave capability as a method for making sense of the change from superposition for an unmistakable state. These models mean to give a deterministic record of quantum estimation.

Quantum Bayesianism (QBism): Quantum Bayesianism, or QBism, centers around the emotional idea of quantum likelihood.

It places that quantum mechanics portrays a specialist's very own convictions about actual frameworks, as opposed to an objective reality.

Every one of these translations offers an interesting viewpoint on quantum peculiarities and addresses the philosophical difficulties presented by quantum mechanics.

6. **Continuous Philosophical Talk:**

The debates encompassing Bohr's translation keep on animating philosophical conversations about the idea of the real world, the job of the spectator, and the ramifications of quantum indeterminacy. These discussions stretch out past mainstream researchers and impact the way of thinking of science and the way of thinking of psyche.

Questions persevere about the connection among awareness and actual reality, the restrictions of human information in figuring out quantum peculiarities, and the idea of logical clarifications even with essential indeterminism.

7. **Influence on Quantum Innovation and Schooling:**

Notwithstanding the discussions and contentions, Bohr's Copenhagen understanding significantly affects the field of quantum innovation. The standards of quantum mechanics, as clarified by the translation, have made ready for the improvement of quantum processing, quantum cryptography, and quantum sensors. These innovations influence the interesting properties of quantum frameworks and can possibly change different fields.

Moreover, the Copenhagen understanding remaining parts a foundation of quantum material science schooling. It fills in as a primary system for showing understudies the basic standards of quantum mechanics, wave-molecule duality, complementarity, and the probabilistic idea of quantum estimations.

Chapter 4

Quantum Experiments
and Paradoxes

The universe of quantum mechanics is a domain of significant secret and perplexing peculiarities, where particles can exist in superpositions, data can be momentarily sent across tremendous distances, and estimations can modify the very properties they look to notice. Quantum examinations and oddities lie at the core of our investigation into this odd and captivating domain, testing our traditional instincts and reshaping how we might interpret the crucial idea of the universe. In this paper, we will leave on an excursion through the most striking quantum trials and Catch 22s, diving into their set of experiences, suggestions, and continuous importance in the realm of material science and reasoning.

1. **Prologue to Quantum Analyses and Oddities**

    Quantum mechanics, which arose in the mid twentieth 100 years, presented an extreme takeoff from old style physical science. While traditional physical science portrayed the way of behaving of plainly visible items with determinism and accuracy, quantum mechanics uncovered an existence where particles displayed wave-molecule duality, where estimations yielded probabilistic results, and where quantum entrapment opposed the limits of reality. These disclosures incited a progression of examinations and mysteries that tested and reshaped how we might interpret reality. Quantum tests are intended to test the way of behaving of quantum substances, like electrons, photons, and particles, under controlled conditions. Oddities emerge when the results of these tests conflict with our traditional instincts and assumptions. Probably the most well known quantum tests and oddities incorporate the twofold cut explore, quantum snare (EPR Catch 22 and Chime's hypothesis), Schrödinger's feline conundrum, and the quantum Zeno impact, among others.

2. **The Twofold Cut Trial: A Preface to Quantum Peculiarity**

The twofold cut try remains as a foundation of quantum material science and fills in as a preface to the unusual and illogical way of behaving of quantum elements. This trial, first performed by Thomas Youthful in the mid nineteenth hundred years and refined

with quantum particles in the twentieth 100 years, includes terminating particles, like electrons or photons, at a hindrance with two cuts.

In the traditional world, one would anticipate that particles should act like minuscule billiard balls, going through one of the cuts and making a basic example on the finder screen behind the obstruction. In any case, in the quantum domain, the outcomes are significantly unique:

Wave-Molecule Duality: Rather than acting as particles, quantum substances show both molecule like and wave-like way of behaving. When unseen, they go through the two cuts at the same time, making an impedance design on the identifier screen, like waves in a lake obstructing one another. This proposes that particles have a wave-like nature, an idea known as wave-molecule duality.

Breakdown of the Wave Capability: When a spectator estimates what cut the molecule goes through, the impedance design vanishes, and the molecule acts as an old style molecule, going through one cut or the other. This is because of the breakdown of the quantum wave capability, which portrays the probabilistic appropriation of the molecule's potential states. The demonstration of estimation falls the wave capability to a solitary, clear result.

The twofold cut try difficulties our traditional instincts about particles having obvious directions. It shows the way that quantum substances can exist in a condition of superposition, where they all the while possess various states, and their way of behaving can change contingent upon whether they are noticed.

3. **Quantum Entrapment: Einstein's Protest and Ringer's Hypothesis**

Quantum entrapment is perhaps of the most bewildering and charming peculiarity in quantum mechanics. It happens when at least two particles become corresponded so that the condition of one molecule is subject to the condition of another, in any event, when isolated by immense distances. Quantum entrapment is at the core of two weighty oddities: the Einstein-Podolsky-Rosen (EPR) mystery and Ringer's hypothesis.

1. **The Einstein-Podolsky-Rosen (EPR) Conundrum:**
   In 1935, Albert Einstein, Boris Podolsky, and Nathan Rosen formed the EPR Catch 22 to challenge the culmination and probabilistic nature of quantum mechanics. The oddity relies on the idea of snare and the properties of two entrapped particles.
   The EPR oddity starts with the formation of a caught sets of particles, like electrons. These particles are made so that their properties, like twist or polarization, are related. In the event that one molecule has a specific twist, the other high priority the contrary twist.
   Here is the mystery: When one molecule is estimated, its state

promptly implodes to an unmistakable result, and because of snare, the condition of the other molecule likewise falls to the contrary result, no matter what the distance between them. This apparently suggests that data or impact ventures promptly between the particles, abusing the rule that no data can proliferate quicker than the speed of light.

Einstein found this part of quantum trap profoundly disturbing, as it seemed to challenge the relativistic idea of causality, which restricts quicker than-light correspondence. He contended that there should be "covered up factors," inconspicuous properties of particles, that could represent the related results without abusing causality.

2. Ringer's Hypothesis and Viewpoint's Tests:

During the 1960s, physicist John Ringer figured out a hypothesis that gave a way to test the expectations of quantum mechanics and secret factors speculations. That's what ringer's hypothesis showed assuming secret factors speculations were right, they would make expectations that varied from those of quantum mechanics.

To test Ringer's hypothesis, physicist Alain Viewpoint led a progression of examinations during the 1980s. The consequences of these trials were shocking: they affirmed the expectations of quantum mechanics and precluded secret factors hypotheses. The analyses exhibited that the connections saw between trapped particles couldn't be made sense of by any old style or deterministic model.

Ringer's hypothesis and Viewpoint's analyses serious areas of strength for offered help for the probabilistic and non-nearby parts of quantum snare. While they didn't determine the philosophical discussions encompassing quantum entrapment, they supported the standards of quantum mechanics and tested traditional instincts.

IV. Schrödinger's Feline Mystery: A Story of Superposition and Estimation

Schrödinger's feline conundrum, proposed by physicist Erwin Schrödinger in 1935, is a psychological study that shows the unusual results of quantum superposition and the job of estimation in quantum mechanics.

In the psychological study, a feline is put in a fixed box with a radioactive iota, a Geiger counter, a vial of toxin, and a mallet. The Geiger counter is associated with the radioactive particle and will set off the arrival of the toxic substance in the event that it distinguishes radiation. As indicated by quantum mechanics:

Quantum Superposition: Before the crate is opened and noticed, the quantum condition of the radioactive iota exists in a superposition, meaning it is at the same time in a rotted and undecayed state. Thusly, the Geiger counter is likewise in a superposition of set off and untriggered states.

Entrapment with the Feline: The feline becomes entrapped with the quantum condition of the particle. In the event that the particle rots and triggers the Geiger counter, the feline is presented to the toxin and kicks the bucket. Assuming the particle remains undecayed, the feline makes due.

The mystery emerges while thinking about the condition of the feline. As per quantum mechanics, the feline is in a superposition of being both alive and dead until the case is opened and noticed. This suggests that, until noticed, the feline exists in a condition of quantum limbo, where it is neither conclusively alive nor absolutely dead.

Schrödinger proposed this Catch 22 as a reductio promotion absurdum, featuring what he saw as the crazy ramifications of quantum mechanics when applied to plainly visible items like felines. It highlights the job of estimation in imploding the quantum superposition and deciding the condition of a framework.

While Schrödinger's feline mystery is in many cases introduced as a psychological study, it brings up significant issues about the limit between the quantum and traditional universes, the job of perception, and the idea of actual reality.

V. The Quantum Zeno Impact: Resisting the Progression of Time

The quantum Zeno impact is an interesting peculiarity that challenges our traditional comprehension of the progression of time and the development of quantum states. It is named after the antiquated Greek thinker Zeno of Elea, who formed oddities about the idea of movement.

In quantum mechanics, the Zeno impact happens when rehashed estimations or perceptions of a quantum framework keep it from developing or changing after some time. Basically, the demonstration of ceaseless estimation "freezes" the quantum state in a specific setup.

To comprehend the quantum Zeno impact, consider a quantum framework, for example, the rot of a radioactive molecule, that has a specific likelihood of progressing starting with one state then onto the next after some time. As per quantum mechanics, in the event that the framework is consistently estimated or noticed, the likelihood of it changing state diminishes. At the end of the day, successive estimations can restrain the framework from going through its normal advancement.

This impact difficulties our old style thought of time as a persistent stream, proposing that in the quantum domain, time can be affected and compelled by the demonstration of perception. The quantum Zeno impact has been tentatively seen in different frameworks, including the rot of shaky particles and the development of quantum turns.

VI. The Deferred Decision Quantum Eraser: Retroactively Deciding the Past

The deferred decision quantum eraser try is a psyche bowing investigation of the retroactive impacts of estimation on the way of behaving of quantum particles. It was intended to research the job of estimation and perception in deciding the previous condition of a quantum framework.

In this examination, a photon is terminated at a shaft splitter, making two snared photons with connected polarizations. One photon goes through a couple of cuts (a twofold cut arrangement) prior to arriving at an indicator screen, while the other photon is sent on an alternate way.

Wave-Molecule Duality: Like the twofold cut explore, the photon that movements through the cuts displays wave-molecule duality, making an impedance design when unseen.

Deferred Decision: Here's where the investigation becomes captivating. The decision of whether to quantify what cut the primary photon goes through is made after the photon has previously gone through the cuts yet before it arrives at the locator screen.

Quantum Eraser: Surprisingly, the decision to quantify or not measure the principal photon retroactively decides if the subsequent photon shows an impedance design or acts like a traditional molecule. In the event that the estimation data is eradicated or "unseen" after the subsequent photon has gone through the cuts however prior to arriving at the locator, the obstruction design reappears.

This examination recommends that the way of behaving of particles in the past can be affected by choices made in the present, bringing up significant issues about causality and the job of perception in deciding the past. It shows that the demonstration of estimation retroactively affects the quantum framework.

VII. The Job of Quantum Analyses and Mysteries in Science and Reasoning

Quantum investigations and Catch 22s act as pots for testing the limits of how we might interpret reality. They challenge our traditional instincts, uncover the weird and strange way of behaving of quantum substances, and feature the significant ramifications of quantum mechanics for the idea of actual reality.

These analyses have extensive outcomes in both science and theory:

Philosophical Ramifications: Quantum examinations and Catch 22s bring up significant philosophical issues about the idea of the real world, the job of the spectator, the idea of determinism versus indeterminism, and the connection between the quantum and old

style universes. They challenge customary originations of causality and the progression of time.

Reasonable Applications: Quantum peculiarities, as uncovered by these analyses, structure the groundwork of quantum innovations, including quantum registering, quantum cryptography, and quantum sensors. These innovations can possibly change fields, for example, data handling, cryptography, and materials science.

Progressing Logical Request: Quantum analyses and conundrums keep on driving logical examination and investigation. They motivate new examinations concerning the key idea of quantum mechanics, the quest for more profound clarifications, and the journey for a brought together hypothesis that can accommodate quantum mechanics with the hypothesis of general relativity.

4.1 Discussion of experimental evidence and paradoxes that challenged classical physics.

Traditional physical science, otherwise called Newtonian physical science, was the predominant system for figuring out the actual world for a really long time. In any case, as logical request advanced into the late nineteenth and mid twentieth hundreds of years, a progression of trial proof and oddities arose that on a very basic level tested the old style material science worldview. These difficulties made ready for the improvement of current physical science, including the hypothesis of relativity and quantum mechanics. In this exposition, we will investigate a portion of the critical exploratory proof and oddities that assumed a vital part in testing old style material science.

1. **Michelson-Morley Investigation: The Invalid Outcome that Changed Physical science**

   Perhaps of the most popular examination throughout the entire existence of material science is the Michelson-Morley explore, directed in 1887 by Albert A. Michelson and Edward W. Morley.

This investigation was intended to distinguish the presence of the luminiferous ether, a medium that was accepted to be essential for the spread of light waves.

In the old style wave hypothesis of light, it was felt that light waves went through the ether, similar as sound waves travel through air. Assuming this were the situation, Earth's movement through space ought to make a recognizable "ether wind" that would influence the speed of light every which way.

To test this speculation, Michelson and Morley set up an inter-ferometer that split a light emission into two opposite ways, with one way lined up with Earth's movement around the Sun and the other opposite to it. As per old style material science, the speed of light in the two bearings ought to have contrasted because of the ether wind.

Notwithstanding, to the shock of established researchers, the Michelson-Morley try yielded an invalid outcome. That is, there was no perceptible distinction in the speed of light in the two headings, recommending that the ether didn't exist or that it didn't have the properties that traditional physical science had credited to it.

The Michelson-Morley analyze tested old style physical science by showing that the traditional idea of the ether was imperfect or nonexistent. This outcome assumed a urgent part in the improvement of Einstein's hypothesis of exceptional relativity, which presented the possibility that the speed of light is a steady for all spectators and is free of the movement of the source or the eyewitness.

2. **Blackbody Radiation and the Bright Fiasco: Quantum In-surgency**

At the turn of the twentieth hundred years, the investigation of blackbody radiation represented a critical test to old style physi-cal science. A blackbody is a romanticized object that ingests all occurrence radiation and discharges radiation across a nonstop range of frequencies.

Traditional physical science, in light of the electromagnetic hypothesis of Maxwell, anticipated that the force of blackbody radiation would increment without bound as the recurrence of radiation expanded, prompting what became known as the "bright disaster." This expectation went against trial perceptions. One of the vital figures in tending to this Catch 22 was Max Planck. In 1900, Planck presented the idea of quantization, recommending that electromagnetic radiation must be produced or retained in discrete bundles or "quanta." He proposed a recipe, presently known as Planck's regulation, that precisely portrayed the unearthly dispersion of blackbody radiation.

Planck's quantization of energy tested old style physical science by presenting the possibility that energy levels were quantized, as opposed to constantly factor, and that the way of behaving of particles and radiation on the nuclear scale was essentially unique in relation to traditional assumptions. This obvious the start of quantum mechanics, another structure that would reform material science.

3. **Photoelectric Impact: Einstein's Quantum Speculation**

The photoelectric impact, first saw by Heinrich Hertz in 1887, includes the discharge of electrons from a material when it is presented to light. Traditional physical science couldn't make sense of the way of behaving of the photoelectric impact, especially as to the connection between the force of light and the active energy of transmitted electrons.

In 1905, Albert Einstein proposed a pivotal clarification of the photoelectric impact in view of his quantum speculation. He proposed that light isn't consistent yet comprises of discrete bundles of energy called "quanta" or photons. These photons have energy relative to their recurrence, as portrayed by Planck's quantum hypothesis.

Einstein's hypothesis of the photoelectric impact effectively made sense of the trial information by suggesting that electrons are launched out from a material when they ingest individual photons with adequate energy to defeat the material's limiting powers. The dynamic energy of the radiated electrons was straightforwardly connected with the energy of the ingested photons.

Einstein's work on the photoelectric impact major areas of strength for gave to the quantum idea of light and tested old style physical science by presenting the idea of the photon as a discrete molecule of energy. This was a pivotal move toward the improvement of quantum mechanics.

4. **Rutherford's Gold Foil Investigation: The Atomic Iota**

Traditional physical science had long maintained the point of view that iotas were inseparable, strong circles with electrons circling around the core. Notwithstanding, Ernest Rutherford's gold foil explore, directed in 1909, gave trial proof that tested this old style model and brought forth the idea of the atomic iota.

In the trial, Rutherford and his partners barraged a slender sheet of gold foil with alpha particles. As indicated by the traditional model, a large portion of the alpha particles ought to have gone through the foil with just minor diversions because of the electron cloud encompassing the iota.

Incredibly, Rutherford's group saw that a few alpha particles were diverted at large angles, and a couple of even bobbed straightforwardly in reverse. This outcome was conflicting with the forecasts of the traditional model and must be made sense of in the event that molecules contained a minuscule, huge core at their middle, which conveyed a positive charge.

Rutherford's gold foil analyze tested traditional physical science by uncovering that the design of the molecule was tremendously not the same as the recently acknowledged model. It presented the possibility that particles were for the most part void space, with electrons circling around a thick, decidedly charged core.

5. **Wave-Molecule Duality: The Twofold Cut Investigation**
The twofold cut explore, which we talked about before with regards to quantum mechanics, additionally tested traditional physical science by showing the wave-molecule duality of particles like electrons and photons.

Wave-Like Way of behaving: When particles were not noticed, they displayed obstruction designs on the locator screen, as though they were waves. This conduct proposed that particles had wave-like properties.

Molecule Like Way of behaving: When the particles were noticed, the obstruction designs vanished, and the particles acted like traditional particles, going through one cut or the other.

The twofold cut try tested traditional material science by showing the way that particles could display both wave-like and molecule like way of behaving, contingent upon whether they were noticed. This duality was in a general sense contrary with old style material science, which portrayed particles as unmistakable, confined elements.

6. **The Lorentz Change and Unique Relativity**

The Michelson-Morley try, alongside other exploratory proof, represented a test to old style material science with respect to the idea of outright existence. The invalid consequence of the Michelson-Morley explore proposed that the speed of light was consistent for all spectators, no matter what their movement.

Because of this test, Albert Einstein fostered the hypothesis of exceptional relativity, which presented the idea of spacetime and the Lorentz change conditions. Unique relativity supplanted old style thoughts of outright existence with the possibility that the laws of physical science were invariant for all inertial onlookers, and that reality were relative.

The hypothesis of extraordinary relativity prompted historic outcomes, including time widening, length constriction, and the renowned condition $E=mc^2$, which related energy and mass. These thoughts tested traditional material science by essentially changing comprehension we might interpret the idea of existence, and they have been affirmed by various trials, like the estimations of the muon's lifetime in high-energy molecule gas pedals.

4.2 The double-slit experiment, wave-particle duality, and the Uncertainty Principle.

The twofold cut analyze is quite possibly of the most notorious and confounding analysis throughout the entire existence of material science. It epitomizes the peculiarity

of wave-molecule duality, a crucial part of quantum mechanics. Besides, the twofold cut try gives a passage to understanding the Heisenberg Vulnerability Rule, a vital idea that supports the inborn constraints of our insight at the quantum level. In this exposition, we will dig into the twofold cut analyze, investigate the significant ramifications of wave-molecule duality, and unwind the complexities of the Vulnerability Standard.

1. **The Twofold Cut Trial: Uncovering Quantum Secret**
1. **The Arrangement:**
   The twofold cut explore includes a basic however reasonably rich arrangement. Envision a hindrance with two restricted cuts through which particles, like electrons or photons, can pass. Past the hindrance, there is a screen that can distinguish the appearance of these particles. From the outset, one could expect that particles would act similar as little billiard balls, going through one of the cuts and making two particular examples on the screen relating to the cuts.
2. **Wave-Like Way of behaving:**
   Nonetheless, the consequences of the twofold cut explore are everything except natural. At the point when a flood of particles is aimed at the hindrance with the two cuts open, something surprising occurs. Rather than delivering two separate examples on the screen, as traditional physical science would foresee, the

particles make an impedance design likened to the example created by waves in a lake.

This impedance design is described by exchanging splendid and dim groups on the screen. The splendid groups compare to districts where the particles have shown up, while the dull groups address locales where they have not. This outcome suggests that particles going through one cut can "meddle" with particles going through the other cut, similar as covering waves produce impedance designs.

3. **Molecule Like Way of behaving:**

The puzzler extends when one thinks about what happens when the trial is altered. On the off chance that an identifier is set at both of the cuts to figure out what cut a molecule goes through, the impedance design vanishes. All things considered, the particles act as though they were old style particles, making two particular examples on the screen.

This perception proposes that the demonstration of estimation or perception implodes the wave-like way of behaving of particles into a molecule like state. The particles are compelled to pick one of the two cuts, and the obstruction design disappears. This is a crucial part of the twofold cut try, featuring the job of the onlooker in characterizing the result of quantum peculiarities.

II. Wave-Molecule Duality: The Core of Quantum Secret

The twofold cut try is a striking exhibit of wave-molecule duality, an idea at the center of quantum mechanics. Wave-molecule duality declares that particles like electrons and photons can show both wave-like and molecule like properties relying upon the states of perception.

1. **Molecule Like Properties:**
   When noticed or estimated, particles act like restricted substances with distinct positions and momenta. They show properties like discrete directions, clear cut positions, and distinct momenta.

This conduct lines up with the old style idea of particles as particular, strong elements.

2. **Wave-Like Properties:**

On the other hand, when particles are not noticed or when they are in a condition of superposition, they show wave-like properties. These properties incorporate obstruction, diffraction, and the capacity to all the while involve various positions. In this state, particles are not bound to explicit ways, and their way of behaving is depicted by likelihood appropriations as opposed to deterministic directions.

Wave-molecule duality challenges our traditional instincts by uncovering that particles can exist in a condition of superposition, possessing numerous positions or momenta all the while. The decision between wave-like and molecule like way of behaving relies upon whether the particles are noticed, which presents a central component of vulnerability into quantum mechanics.

3. **Ramifications of Wave-Molecule Duality:**

The presence of wave-molecule duality has significant ramifications for how we might interpret reality at the quantum level. It highlights the restrictions of traditional physical science in depicting the way of behaving of particles on a limited scale. Besides, it proposes that quantum elements exist in a condition of possibility, with their properties not set in stone until estimated.

Wave-molecule duality likewise challenges the traditional idea of causality, as particles in superposition don't follow deterministic ways yet rather exist in a probabilistic domain. The job of the eyewitness in characterizing the result of quantum occasions accentuates the significant association between the demonstration of estimation and the way of behaving of quantum elements.

III. The Heisenberg Vulnerability Standard: The Limits of Information

Wave-molecule duality is personally associated with the Heisenberg Vulnerability Standard, a central idea in quantum mechanics figured out by Werner Heisenberg in 1927. The Vulnerability Standard puts essential cutoff points on our capacity to all the while know specific sets of corresponding properties of a quantum element, like position and force.

1. **Explanation of the Vulnerability Rule:**
   (h-bar) is the diminished Planck's consistent, around 1.0545718 x 10^-34 J·s.
   In less difficult terms, the Vulnerability Standard expresses that the result of the vulnerability ready (Δx) and the vulnerability in energy (Δp) of a quantum element should continuously be more noteworthy than or equivalent to a steady worth, which is a key consistent of nature.

2. **Ramifications of the Vulnerability Guideline:**
   The Vulnerability Standard suggests that the more definitively we know the place of a quantum substance, the less exactly we can know its force, as well as the other way around. This presents a major limit on our capacity to gauge specific sets of properties with inconsistent accuracy at the same time.
   For instance, with regards to the twofold cut try, on the off chance that we exactly decide the place of a molecule as it goes through one of the cuts, the vulnerability in its energy turns out to be extremely enormous. Thus, we can't foresee with conviction where the molecule will arrive on the screen, as its force vulnerability prompts a spread in potential results.
   The Vulnerability Guideline challenges the old style thought that it is feasible to quantify the position and energy of a molecule with erratic accuracy at the same time. It highlights the intrinsic furthest reaches of our insight at the quantum level and presents a component of principal vulnerability into the way of behaving of quantum substances.

3. **Past Position and Energy: General Detailing:**

The Heisenberg Vulnerability Standard is frequently connected with the vulnerability ready and force, however it has a more broad plan that applies to sets of reciprocal observables. These observables can incorporate energy and time, precise force parts, and other actual properties.

IV. Quantum Mechanics and the Idea of The real world

The twofold cut try, wave-molecule duality, and the Heisenberg Vulnerability Rule all in all challenge our traditional instincts about the idea of the real world. They portray the quantum world that is significantly not quite the same as the deterministic, old style universe imagined by early physicists like Isaac Newton.

1. **Probabilistic Nature:**

   In the quantum domain, results are intrinsically probabilistic. Wave capabilities, which portray the quantum conditions of particles, give us probabilities of finding a molecule in a specific state when estimated. This probabilistic nature of quantum mechanics is a takeoff from the deterministic directions anticipated by traditional material science.

2. Spectator Subordinate Reality:

   The job of the eyewitness in quantum mechanics is a wellspring of philosophical discussion and interest. The demonstration of estimation or perception implodes the wave capability, deciding the result of a trial. This proposes that reality at the quantum level is onlooker subordinate, and the demonstration of estimation assumes a focal part in characterizing what is genuine.

3. **Complementarity:**

   The idea of complementarity, presented by Niels Bohr, is firmly connected with wave-molecule duality and the Vulnerability Rule. It places that specific sets of correlative properties can't be all the while estimated with inconsistent accuracy. For instance, on the off chance that we know the place of a molecule with high accuracy, we lose data about its energy, as well as the other way around.

4. **Quantum Superposition:**
Wave-molecule duality likewise prompts the possibility of quantum superposition, where particles can exist in different states all the while until estimated. This idea has reasonable applications in quantum figuring, where quantum bits (qubits) can address numerous states immediately, empowering quantum PCs to tackle specific issues dramatically quicker than old style PCs.

5. **Difficulties to Instinct:**

The peculiarities depicted by the twofold cut try, wave-molecule duality, and the Vulnerability Standard test our traditional instincts about the way of behaving of actual frameworks. They uncover the impediments of traditional material science when applied to the quantum domain and underline the requirement for another structure to depict the way of behaving of particles for little scopes.

4.3 How these experiments contributed to the clash between Einstein and Bohr.

The conflict between Albert Einstein and Niels Bohr addresses perhaps of the most well known and getting through banter throughout the entire existence of physical science. At its heart were significant conflicts over the understanding of quantum mechanics and the idea of actual reality. This conflict was filled, to a great extent, by the noteworthy examinations and disclosures examined before, like the twofold cut try, wave-molecule duality, and the Heisenberg Vulnerability Standard. In this article, we will investigate how these tests added to the conflict among Einstein and Bohr, revealing insight into their different perspectives and the persevering through effect of their discussion on the groundworks of quantum hypothesis.

1. **The Development of Quantum Mechanics**

   To comprehend the conflict among Einstein and Bohr, it is fundamental to see the value in the setting in which quantum mechanics arose. In the mid twentieth hundred years, physicists were wrestling with the baffling way of behaving of particles at the nuclear and subatomic levels. Traditional physical science, which had been exceptionally fruitful in portraying the naturally visible world, wavered when applied to these little scopes.

   The trials talked about in past segments, especially the twofold cut explore and the Heisenberg Vulnerability Standard, broke traditional instincts. They uncovered that particles could display both wave-like and molecule like properties, that the demonstration of estimation on a very basic level affected results, and that specific sets of correlative properties couldn't be all the while estimated with erratic accuracy.

2. **Einstein's Incredulity: "God Doesn't Play Dice"**

   Albert Einstein, one of the best physicists of the twentieth 100 years, was profoundly doubtful of the probabilistic and indeterministic nature of quantum mechanics. He broadly jested, "God doesn't play dice with the universe," communicating his conviction that the universe ought with comply to deterministic regulations, similar as those depicted by old style material science.

   Einstein's wariness was attached in his obligation to authenticity and determinism. He accepted that there were basic, stowed away factors that could completely depict the way of behaving of quantum frameworks. As far as he might be concerned, the clear arbitrariness and capriciousness of quantum mechanics were a fragmented depiction of the real world.

3. **Bohr's Copenhagen Understanding: Embracing Quantum Probabilities**

   On the opposite side of the discussion stood Niels Bohr, a Danish physicist who created what is currently known as the Copenhagen translation of quantum mechanics. Bohr embraced the probabilistic and indeterministic nature of quantum hypothesis, contending that it was a finished and precise portrayal of the quantum world.

Integral to Bohr's translation was the possibility of complementarity, which he presented. Complementarity proposed that specific sets of corresponding properties, like position and force, couldn't be at the same time known with inconsistent accuracy. Bohr accepted that traditional ideas of determinism and causality didn't matter at the quantum level, and that we needed to acknowledge the intrinsic constraints of our insight.

4. **The Conflict Over the EPR Catch 22**

The conflict among Einstein and Bohr reached a critical stage with the distribution of the renowned EPR paper in 1935. The EPR paper, wrote by Einstein, Boris Podolsky, and Nathan Rosen, introduced a psychological study expected to feature what they saw as an imperfection in the groundworks of quantum mechanics. This psychological test has since become known as the EPR mystery.

The EPR contention depended on the possibility of "ensnarement," a quantum peculiarity wherein two particles become connected so that estimating one molecule immediately decides the properties of the other, no matter what the distance between them. Einstein, Podolsky, and that's what rosen contended assuming quantum mechanics were right, it would infer "creepy activity a ways off," disregarding the standards of territory and authenticity.

Einstein, as he continued looking for determinism and authenticity, accepted that quantum mechanics was deficient and that it should contain stowed away factors that decided the results of estimations. He utilized the EPR conundrum to challenge the culmination and understanding of quantum mechanics.

5. **Bohr's Reaction: Non-Nearby Real factors**

Niels Bohr, in light of the EPR conundrum, shielded the

standards of quantum mechanics and embraced the non-nearby and probabilistic nature of the hypothesis. He contended that ensnarement didn't abuse the standards of territory and authenticity as Einstein guaranteed.

Bohr's key understanding was that quantum frameworks didn't have clear properties until estimated, and these properties became corresponded promptly when entrapped particles were estimated. He contended that the demonstration of estimation didn't include the exchange of actual data or "activity a ways off" in the traditional sense. All

things being equal, it mirrored the non-neighborhood interconnectedness of quantum substances.

Bohr's reaction to the EPR conundrum was to maintain the probabilistic and non-nearby nature of quantum mechanics and to dismiss the thought of stowed away factors. He trusted that the quantum world had its own exceptional arrangement of rules and couldn't be figured out utilizing traditional instincts.

6. **The Chime's Hypothesis Trials: Testing Quantum Ensnarement**

The conflict among Einstein and Bohr stayed unsettled until the 1960s when physicist John Chime planned a numerical hypothesis, known as Ringer's hypothesis, that gave a method for testing the expectations of quantum ensnarement tentatively. That's what chime's hypothesis showed, assuming quantum mechanics was right, certain connections between's estimations on trapped particles couldn't be made sense of by any hypothesis including stowed away factors.

In the many years that followed, a progression of trials were led to test the expectations of Ringer's hypothesis. These examinations, for example, the Perspective analysis during the 1980s, reliably upheld the quantum mechanical expectations of ensnarement and non-region, precluding the chance of stowed away factors.

Chime's hypothesis tests areas of strength for gave for Bohr's translation of quantum mechanics, supporting the probabilistic and non-nearby nature of the hypothesis. They managed a critical disaster for Einstein's vision of stowed away factors and determinism.

7. The Tradition of the Einstein-Bohr Discussion

The conflict among Einstein and Bohr was not only a scholastic conflict but rather a central discussion about the idea of actual reality. It uncovered profound philosophical contrasts in their ways to deal with grasping the quantum world.

1. **Einstein's Inheritance: Stowed away Factors and Authenticity**
   Einstein's incredulity about the fulfillment of quantum mechanics and his journey for buried factors left an enduring inheritance. His quest for a deterministic and sensible hypothesis tested the overarching translations of quantum mechanics. While stowed away factors speculations have to a great extent been precluded by Chime's hypothesis explores, Einstein's obligation to authenticity keeps on impacting conversations about the underpinnings of quantum hypothesis.

2. Bohr's Heritage: Complementarity and Quantum Translation
   Niels Bohr's Copenhagen understanding, with its hug of complementarity, stays a foundation of quantum mechanics. It has formed the manner in which physicists contemplate quantum peculiarities and the restrictions of our insight at the quantum level. Bohr's non-neighborhood and probabilistic perspective has persevered as an essential part of quantum hypothesis.

3. **Goal of the Discussion: Non-Traditional Real factors**

Eventually, the conflict among Einstein and Bohr, while serious and logically significant, didn't bring about a conclusive triumph for one or the other side. All things considered, it featured the non-traditional nature of the quantum world, where old style instincts and deterministic speculations don't make a difference.

The tradition of this discussion has highlighted the wealth and intricacy of quantum mechanics. It has additionally stressed the significance of exploratory proof in molding how we might interpret the quantum domain. Quantum mechanics, with its probabilistic and non-neighborhood highlights, keeps on testing our instincts about the idea of the real world, welcoming us to wrestle with its secrets and logical inconsistencies.

5

Chapter 5

The Solvay Conferences

The Solvay Meetings, a progression of select social events of the world's most splendid personalities in material science, play had a critical impact in molding the course of present day physical science. These meetings have been a gathering for the trading of notable thoughts, warmed discusses, and the improvement of quantum mechanics, relativity hypothesis, and other central ideas. In this exhaustive paper, we will investigate the set of experiences, importance, and effect of the Solvay Gatherings on the field of physical science.

1. **Prologue to the Solvay Gatherings**
   The Solvay Gatherings are a progression of greeting just gatherings of physicists that have been held roughly like clockwork since the mid twentieth 100 years. These meetings, named after the Belgian scientific expert and industrialist Ernest Solvay, have united the absolute most unmistakable researchers of their opportunity to examine and discuss central questions in physical science. The Solvay Meetings have been instrumental in propelling the comprehension of quantum mechanics, the hypothesis of relativity, and the idea of the subatomic world.

2. **The Early Years: The Introduction of Quantum Mechanics**

The primary Solvay Meeting occurred in 1911 in Brussels, Belgium. This debut gathering was coordinated by Ernest Solvay himself, and it denoted the start of a custom that would go on for north of 100 years. The gathering zeroed in on the arising field of nuclear and sub-atomic material science, which was encountering a time of significant change.

Eminent participants at the primary Solvay Meeting included Albert Einstein, Max Planck, and Marie Curie, among others. It was at this gathering that the renowned Solvay Solids photo was taken, catching a gathering of illuminating presences who might proceed to make notable commitments to science.

The 1911 Solvay Meeting established the groundwork for the advancement of quantum mechanics. During this gathering, Max Planck introduced his thoughts on the quantization of energy, which had previously altered the comprehension of blackbody radiation. Albert Einstein, who had as of late presented the hypothesis of the photoelectric impact, likewise made huge commitments to the conversations. These early discussions and introductions set up for the transformation in material science that was to come.

3. **The 1927 Solvay Meeting: The Introduction of Quantum Mechanics**

One of the most notable Solvay Meetings happened in 1927, likewise in Brussels. This meeting is frequently alluded to as the "Fifth Solvay Gathering" and is commended for its vital job in the advancement of quantum mechanics.

At the 1927 Solvay Meeting, probably the most brilliant personalities in material science accumulated to talk about the significant ramifications of quantum mechanics. Participants included Albert Einstein, Niels Bohr, Werner Heisenberg, Erwin Schrödinger, Max Conceived, and numerous others. It was a genuine social occasion of the initial architects of quantum hypothesis.

The gathering conversations were serious, with energetic discussions among Einstein and Bohr at the middle. Albert Einstein broadly expressed, "God doesn't play dice with the universe," communicating his distress with the probabilistic and indeterministic nature of quantum mechanics. Niels Bohr, then again, supported the Copenhagen translation, stressing the job of perception and the constraints of traditional instincts at the quantum level.

The 1927 Solvay Meeting additionally created the popular Solvay Gathering photo, highlighting the illuminators of quantum mechanics, which has turned into a notable picture throughout the entire existence of science.

4. **Quantum Mechanics and Then some: Resulting Solvay Gatherings**

The Solvay Meetings kept on being a gathering for conversations on the turn of events and translation of quantum mechanics in the ensuing many years. While the early meetings zeroed in principally on quantum hypothesis, later social occasions tended to many points in material science, including atomic physical science, measurable mechanics, and the construction of issue.

One prominent Solvay Gathering, held in 1933, tended to the subject of electrons and photons. This gathering remembered conversations for the arising field of quantum electrodynamics (QED), which tried to accommodate quantum mechanics with the hypothesis of electromagnetism. Eminent participants included Paul Dirac and Wolfgang Pauli.

The 1954 Solvay Meeting, frequently alluded to as the "Eighth Solvay Gathering," was a critical occasion throughout the entire existence of physical science. It zeroed in on the principal connections of rudimentary particles and laid the preparation for the improvement of the Standard Model of molecule material science, which portrays the electromagnetic, powerless, and solid atomic powers.

The Solvay Gatherings kept on developing, mirroring the

changing scene of material science. They assumed a pivotal part in cultivating cooperation, sharing of thoughts, and the improvement of new speculations and examinations.

5. **The Job of the Solvay Gatherings in Forming Material science**

The Solvay Gatherings significantly affect the field of physical science, impacting the heading of exploration and the advancement of key hypotheses. Here are a portion of the manners by which the Solvay Gatherings have molded material science:

1. **Headway of Quantum Mechanics:**
   The early Solvay Gatherings assumed a focal part in the improvement of quantum mechanics. They gave a stage to the trading of thoughts among illuminating presences like Planck, Einstein, Bohr, and Schrödinger, who all in all established the groundwork for the new quantum hypothesis. The discussions and conversations at these gatherings refined the ideas of quantum mechanics and added to its inescapable acknowledgment.

2. **Goal of Logical Conflicts:**
   The Solvay Meetings were social affairs of extraordinary personalities as well as fields for logical discussion and goal of conflicts. The energetic discussions among Einstein and Bohr at the 1927 gathering, for instance, explained the standards of quantum mechanics and the idea of reality at the quantum level. While the discussions were frequently warmed, they were fundamental for refining and cementing the speculations being talked about.

3. **Encouraging Coordinated effort:**
   The Solvay Meetings united physicists from around the world, setting out open doors for coordinated effort and cross-preparation of thoughts. The collaborations among researchers with assorted foundations and points of view prompted new experiences and forward leaps in different areas of material science.

These meetings worked with the improvement of a worldwide academic local area.

4. **Setting Exploration Plans:**
The Solvay Meetings frequently set the plan for future exploration in physical science. Subjects talked about at these gatherings were significant by their own doing as well as impacted the course of logical request. For example, the 1954 gathering on rudimentary

particles characterized the exploration program that in the long run prompted the improvement of the Standard Model of molecule material science.

5. **Advancing Interdisciplinary Methodologies:**

The Solvay Meetings urged interdisciplinary ways to deal with physical science. Participants from various subfields of physical science met up to talk about normal issues and difficulties. This interdisciplinary trade of thoughts has been instrumental in propelling comprehension we might interpret the actual universe.

VI. The Tradition of the Solvay Meetings

The tradition of the Solvay Meetings perseveres in the chronicles of material science history. These meetings not just high level comprehension we might interpret the quantum world yet additionally exhibited the force of coordinated effort and open logical talk. Their impact stretches out a long ways past the meeting rooms where they were held.

1. **Propelling Central Physical science:**
The Solvay Gatherings have been instrumental in propelling comprehension we might interpret essential material science. They gave a stage to the improvement of quantum mechanics, the hypothesis of relativity, and the underpinnings of current molecule material science. The bits of knowledge and leap forwards that arose out of these meetings have molded the scene of material science research for a really long time.

2. **Forming the Course of Logical Revelation:**
   The discussions and conversations at the Solvay Gatherings helped shape the course of logical revelation in the twentieth 100 years. They directed the course of examination, distinguished basic inquiries, and persuaded researchers to investigate new outskirts. The meetings assumed a urgent part in setting the exploration plan for ages of physicists.

3. **Rousing People in the future:**

The Solvay Meetings keep on motivating people in the future of physicists. The narratives of energetic discussions and pivotal disclosures act as a wake up call of the extraordinary force of logical coordinated effort and scholarly trade. The Solvay Meetings embody the quest for information and the steady mission to grasp the laws of the universe.

5.1 Exploration of the famous Solvay Conferences where Einstein and Bohr debated.

The Solvay Gatherings stand as notable occasions throughout the entire existence of science, giving a phase to the world's superior physicists to meet, discussion, and push the limits of how we might interpret the actual universe. Among the various lights who graced these social events, the discussions between Albert Einstein and Niels Bohr at the Solvay Gatherings of the 1920s and 1930s hold a unique spot in the chronicles of material science. In this investigation, we dig into the popular Solvay Meetings where Einstein and Bohr took part in warmed conversations that would essentially shape the course of present day material science.

1. **The Solvay Meetings: A Brief look into History**
   The Solvay Meetings, named after Belgian scientific expert and industrialist Ernest Solvay, started in 1911 and have proceeded to the current day, yet with interferences because of worldwide occasions like Universal Conflicts. These gatherings united the absolute most brilliant personalities in physical science and science to

examine basic inquiries in the field. The early gatherings zeroed in basically on the expanding field of quantum mechanics, while later versions covered subjects going from atomic material science to the crucial powers of nature.

2. **The Setting of Quantum Mechanics**

To comprehend the meaning of the discussions among Einstein and Bohr at the Solvay Meetings, we should get a handle on the scholarly milieu of the time. At the turn of the twentieth hundred years, old style material science, essentially portrayed by the laws of Newton and Maxwell, had ruled for quite a long time. In any case, as researchers tested further into the nuclear and subatomic domains, they experienced peculiarities that old style physical science couldn't make sense of.

In the late nineteenth and mid twentieth hundreds of years, tests like the photoelectric impact, blackbody radiation, and the way of behaving of electrons in molecules challenged traditional expectations. It was in this setting that the beginning field of quantum mechanics was conceived, with pioneers like Max Planck, Albert Einstein, and Niels Bohr driving the way.

3. **The 1927 Solvay Gathering: Quantum Mechanics Becomes the overwhelming focus**

The 1927 Solvay Gathering, frequently alluded to as the "Fifth Solvay Meeting," was a turning point throughout the entire existence of physical science. Held in Brussels, this gathering was devoted to talking about the new and puzzling hypothesis of quantum mechanics. The lights in participation included Albert Einstein, Niels Bohr, Max Planck,

Werner Heisenberg, Erwin Schrödinger, Max Conceived, and Louis de Broglie, among others.

The discussions at this gathering, especially those among Einstein and Bohr, would resonate for quite a long time. The focal issue under a microscope was the translation of quantum mechanics' and, specifically, the job of likelihood and determinism in the quantum world.

4. **Einstein's Wariness: "God Doesn't Play Dice"**

At the core of Einstein's wariness was his obligation to a deterministic and pragmatist perspective. He broadly commented, "God doesn't play dice with the universe," communicating his distress with the inborn probabilistic and indeterministic nature of quantum mechanics.

Einstein's suspicion was established in his conviction that there should exist stowed away factors — obscure variables that would decide the results of quantum occasions. He felt that the clear arbitrariness and vulnerability of quantum mechanics were a deficient portrayal of actual reality.

5. **Bohr's Copenhagen Understanding: Embracing Quantum Probabilities**

Niels Bohr, then again, supported the Copenhagen understanding of quantum mechanics. He contended that the probabilistic and indeterministic nature of quantum hypothesis was not an inadequacy but rather a principal element of the quantum world.

Vital to Bohr's understanding was the idea of complementarity, which set that specific sets of corresponding properties, like position and force, couldn't be all the while estimated with erratic accuracy. Bohr accepted that old style ideas of determinism and causality didn't have any significant bearing at the quantum level, and that we needed to acknowledge the intrinsic restrictions of our insight.

6. **The Discussion at the 1927 Solvay Meeting**

The conversations at the 1927 Solvay Meeting were extreme, with enthusiastic discussions among Einstein and Bohr at the middle. Einstein's protests fixated on the possibility that quantum mechanics, as it was perceived at that point, was inadequate. He accepted that a total hypothesis ought to consider the exact forecast of results, not simply probabilistic proclamations.

Bohr, accordingly, contended that quantum mechanics was a finished and independent hypothesis. He fought that the demonstration of estimation was a basic piece of quantum reality and that the indeterminacy and probabilistic nature of quantum mechanics were not because of restrictions in our insight however were central parts of the quantum world.

The discussions among Einstein and Bohr were portrayed by profound philosophical contrasts. Einstein looked to hold an old style, deterministic perspective, while Bohr embraced the probabilistic and indeterministic nature of quantum mechanics. The conflict of thoughts at the 1927 Solvay Gathering didn't bring about a goal, however it set up for additional investigation and refinement of quantum hypothesis.

7. **Ensuing Solvay Meetings and Proceeding with Discussions**

The discussions among Einstein and Bohr at the 1927 Solvay Meeting were only one part in a continuous talk that traversed a few Solvay Gatherings. Ensuing meetings kept on investigating the establishments and ramifications of quantum mechanics, with illuminating presences like Wolfgang Pauli, Paul Dirac, and Richard Feynman joining the conversations.

One outstanding Solvay Meeting, held in 1933, zeroed in on electrons and photons. This gathering tended to the arising field of quantum electrodynamics (QED), which looked to accommodate quantum mechanics with the hypothesis of electromagnetism. The conversations and discussions at this meeting laid the foundation for the advancement of QED.

8. **Goal of the Discussion: Chime's Hypothesis Analyses**

The conflict among Einstein and Bohr stayed unsettled until the 1960s when physicist John Chime formed a numerical hypothesis, known as Ringer's hypothesis, that gave a method for testing the expectations of quantum snare tentatively. That's what chime's hypothesis showed assuming quantum mechanics was right, certain relationships between's estimations on trapped particles couldn't be made sense of by any hypothesis including stowed away factors.

In the many years that followed, a progression of examinations were directed to test the expectations of Chime's hypothesis. These analyses, for example, the Viewpoint explore during the 1980s, reliably upheld the quantum mechanical expectations of snare and non-region, precluding the chance of stowed away factors.

Chime's hypothesis tests areas of strength for gave for Bohr's understanding of quantum mechanics, supporting the probabilistic and non-neighborhood nature of the hypothesis. They managed a critical catastrophe for Einstein's vision of stowed away factors and determinism.

9. The Tradition of the Solvay Gatherings: Molding Present day Material science

The Solvay Meetings and the discussions among Einstein and Bohr have passed on a persevering through inheritance that keeps on forming the field of physical science.

Their importance stretches out past the particular conflicts between these two titans of science.

1. **Propelling Quantum Mechanics:**
   The Solvay Meetings assumed a focal part in propelling comprehension we might interpret quantum mechanics. They gave a stage to the trading of thoughts among lights like Planck, Einstein, Bohr, and Schrödinger, who all in all established the groundwork for the new quantum hypothesis. The discussions and conversations at these meetings refined the ideas of quantum mechanics and added to its far reaching acknowledgment.

2. **Settling Logical Conflicts:**
   The Solvay Meetings were get-togethers of extraordinary personalities as well as fields for logical discussion and goal of conflicts.

The energetic discussions among Einstein and Bohr explained the standards of quantum mechanics and the idea of reality at the quantum level. While the discussions were frequently warmed, they were fundamental for refining and cementing the speculations being talked about.

3. **Encouraging Cooperation:**

The Solvay Meetings united physicists from around the world, setting out open doors for cooperation and cross-preparation of thoughts. The communications among researchers with assorted foundations and points of view prompted new experiences and leap forwards in different areas of physical science. These meetings worked with the improvement of a worldwide academic local area.

4. **Setting Exploration Plans:**

The Solvay Meetings frequently set the plan for future exploration in physical science. Subjects examined at these gatherings were significant by their own doing as well as impacted the bearing of logical request. For example, the 1954 meeting on rudimentary particles characterized the examination program that in the long run prompted the improvement of the Standard Model of molecule physical science.

5. **Advancing Interdisciplinary Methodologies:**

The Solvay Meetings urged interdisciplinary ways to deal with material science. Participants from various subfields of physical science met up to examine normal issues and difficulties. This interdisciplinary trade of thoughts has been instrumental in propelling comprehension we might interpret the actual universe.

5.2 Key moments and debates from the conferences, including their disagreements.

The Solvay Meetings, a progression of social events of the world's most noticeable physicists, have seen essential minutes and discussions that significantly formed the course of present day material science.

Among the most important are the conflicts between Albert Einstein and Niels Bohr, two goliaths of twentieth century physical science. In this investigation, we dive into the critical minutes and discussions from the Solvay Meetings, featuring their conflicts and the persevering through influence on how we might interpret the universe.

1. **The Introduction of Quantum Mechanics: The 1911 Solvay Meeting**

   The primary Solvay Meeting in 1911 denoted a huge achievement throughout the entire existence of material science. This social event, coordinated by Ernest Solvay in Brussels, was devoted to the arising field of nuclear and sub-atomic material science. Striking participants included Max Planck, Marie Curie, Albert Einstein, and Ernest Rutherford, among others.

   Key Minutes and Discussions:

   Planck's Quantum Speculation: Max Planck introduced his earth shattering quantum theory, which presented quantized energy levels. He suggested that energy must be produced or consumed in discrete, quantized units or "quanta." This speculation established the groundwork for quantum mechanics.

   Einstein's Photon Hypothesis: Albert Einstein introduced his hypothesis of the photoelectric impact, which placed that light comprises of discrete bundles of energy called "photons." This hypothesis gave unquestionable proof to the quantization of light and procured Einstein the Nobel Prize in Physical science in 1921.

   Banter on Quantum Authenticity: Albeit not a head on a showdown, the 1911 Solvay Gathering set up for future discussions on the translation of quantum mechanics. While Planck's and Einstein's commitments were significant, they held varying perspectives on the ontological idea of quantum peculiarities.

2. **The 1927 Solvay Gathering: The Conflict of Titans**

   The 1927 Solvay Gathering, frequently alluded to as the "Fifth

Solvay Meeting," stays perhaps of the most famous crossroads throughout the entire existence of material science. Held in Brussels, it united illuminating presences like Albert Einstein, Niels Bohr, Werner Heisenberg, Erwin Schrödinger, Max Conceived, and Louis de Broglie.

Key Minutes and Discussions:

Einstein versus Bohr: The focal subject of the 1927 meeting was the understanding of quantum mechanics, especially the job of likelihood and determinism in the quantum world. Albert Einstein, a steadfast pragmatist, took part in serious discussions with Niels Bohr, who supported the probabilistic Copenhagen translation.

Einstein's Evaluate: Einstein's wariness focused on his well known proclamation, "God doesn't play dice with the universe." He protested the probabilistic idea of quantum mechanics and accepted that a total hypothesis ought to consider deterministic expectations. He looked to challenge the indeterminacy and arbitrariness inborn in quantum mechanics.

Bohr's Copenhagen Understanding: Niels Bohr protected the Copenhagen translation, contending that the probabilistic and indeterministic nature of quantum mechanics was not a constraint but rather an essential component of the quantum world. He underscored the job of perception and estimation in characterizing actual reality at the quantum level.

Complementarity: Bohr presented the idea of complementarity, which placed that specific sets of corresponding properties, like position and energy, couldn't be all the while estimated with inconsistent accuracy. This idea was integral to his translation of quantum mechanics.

The 1927 Solvay Gathering was set apart by enthusiastic discussions and significant philosophical contrasts among Einstein and Bohr. While the gathering didn't bring about a conclusive goal, it left an enduring effect on the turn of events and translation of quantum mechanics.

3. **Ensuing Solvay Meetings: Proceeding with Discussions and Investigations**

The Solvay Gatherings kept on being gatherings for continuous conversations and discussions in physical science. While the conflicts among Einstein and Bohr at the 1927 meeting were fundamental, resulting social occasions tended to different parts of quantum mechanics and the groundworks of material science.

Key Minutes and Discussions:

1933 Solvay Gathering: This meeting zeroed in on electrons and photons and dug into the arising field of quantum electrodynamics (QED). Conversations rotated around the quantization of the electromagnetic field and the compromise of quantum mechanics with electromagnetism.

1948 Solvay Meeting: Held after The Second Great War, this gathering tended to quantum electrodynamics and quantum field hypothesis. It investigated the numerical and calculated difficulties of binding together quantum mechanics with the hypothesis of fields.

1954 Solvay Meeting: Frequently alluded to as the "Eighth Solvay Gathering," this social affair was committed to the essential cooperations of rudimentary particles. Conversations laid the basis for the advancement of the Standard Model of molecule material science, which portrays the electromagnetic, feeble, and solid atomic powers.

4. **Goal of the Discussion: Chime's Hypothesis and Analyses**

The discussions among Einstein and Bohr endured for a really long time, testing the understanding of quantum mechanics. It was only after the definition of John Ringer's hypothesis during the 1960s that a way to tentatively test the idea of quantum ensnarement arose.

Key Minutes and Discussions:

Chime's Hypothesis: John Ringer's hypothesis, distributed in 1964, gave a numerical structure to test the expectations of quantum mechanics with respect to ensnared particles. That's what chime showed assuming quantum mechanics was right, certain connections between's estimations on caught particles couldn't be made sense of by any hypothesis including stowed away factors.

Ringer's Imbalance Trials: In the next many years, a progression of examinations, for example, the Perspective analysis during the 1980s, were directed to test Chime's disparities. These tests reliably upheld the quantum mechanical expectations of snare and non-region, precluding the chance of stowed away factors.

Goal of the Discussion: Ringer's hypothesis tests major areas of strength for gave for Bohr's understanding of quantum mechanics, supporting the probabilistic and non-nearby nature of the hypothesis. They managed a huge disaster for Einstein's vision of stowed away factors and determinism.

5.3 The influence of these conferences on the development of quantum physics.

The Solvay Meetings, a progression of social events of recognized physicists that started in 1911, play had a urgent impact in molding the improvement of quantum material science. These gatherings united probably the best personalities in the field, giving a stage to conversations, discusses, and the trading of historic thoughts. In this

investigation, we will dig into the significant impact of the Solvay Gatherings on the development of quantum physical science.

1. **The Development of Quantum Material science**

   To see the value in the effect of the Solvay Gatherings, it's fundamental to comprehend the setting where they arose. At the turn of the twentieth hundred years, traditional material science, as figured out by Newton, Maxwell, and others, had given an exhaustive structure to grasping the actual world. In any case, as researchers examined further into the idea of issue and radiation, they experienced peculiarities that traditional material science couldn't make sense of.

The late nineteenth and mid twentieth hundreds of years saw a progression of exploratory disclosures and hypothetical improvements that laid the preparation for quantum material science:

Planck's Quantum Speculation (1900): Max Planck's spearheading work on blackbody radiation presented the idea of quantization of energy. He recommended that energy is radiated or consumed in discrete units, or "quanta," as opposed to constantly. This thought denoted the introduction of quantum hypothesis.

Einstein's Hypothesis of the Photoelectric Impact (1905): In his 1905 paper on the photoelectric impact, Albert Einstein recommended that light comprises of discrete bundles of energy called "photons." This hypothesis gave solid proof to the quantization of light and had significant ramifications for the advancement of quantum mechanics.

Bohr's Model of the Hydrogen Molecule (1913): Niels Bohr's model of the hydrogen particle presented the idea of quantized electron circles, where electrons could possess explicit energy levels. This model effectively made sense of the phantom lines of hydrogen and denoted a critical stage towards quantum mechanics.

2. **The Introduction of the Solvay Gatherings**

The principal Solvay Gathering, held in 1911 in Brussels, Belgium, denoted the start of a custom that proceeds right up to the present day. Ernest Solvay, a Belgian scientific expert and industrialist, offered the monetary help for these meetings, which would accumulate driving researchers to examine basic subjects in physical science and science.

The 1911 Solvay Meeting, zeroed in on nuclear and sub-atomic physical science, highlighted striking participants, including Max Planck, Marie Curie, Albert Einstein, and Ernest Rutherford. It was at this meeting that the groundworks of quantum physical science started to come to fruition.

3. **Propelling Quantum Mechanics: Key Minutes from the Solvay Gatherings**

The Solvay Gatherings assumed an essential part in the improvement of quantum mechanics. Key minutes and conversations at these meetings progressed how we might interpret the quantum world:

1. **The 1927 Solvay Meeting: The Conflict of Translations:**
   Einstein versus Bohr: The 1927 Solvay Meeting, frequently alluded to as the "Fifth Solvay Gathering," highlighted extraordinary discussions between Albert Einstein and Niels Bohr. At the core of the conflict was the understanding of quantum mechanics, explicitly the job of likelihood and determinism in the quantum world.
   Einstein's Distrust: Einstein broadly proclaimed, "God doesn't play dice with the universe," communicating his uneasiness with the probabilistic and indeterministic nature of quantum mechanics. He tried to challenge the possibility that quantum occasions were innately arbitrary.
   Bohr's Copenhagen Understanding: Niels Bohr guarded the Copenhagen translation, which stressed the probabilistic and indeterministic nature of quantum mechanics. He contended that the demonstration of estimation was an indispensable piece of quantum reality and that traditional ideas of determinism didn't make a difference at the quantum level.

2. **Ensuing Solvay Meetings: Refining Quantum Mechanics:**

The Solvay Meetings that followed the 1927 social event kept on investigating the establishments and ramifications of quantum mechanics. These gatherings worked with conversations on points, for example, quantum electrodynamics (QED), the quantization of fields, and the crucial communications of rudimentary particles.

IV. Headways in Quantum Electrodynamics (QED)

1933 Solvay Gathering: This meeting zeroed in on electrons and photons, tending to the quantization of the electromagnetic field. Conversations focused on the arising field of quantum electrodynamics (QED), which expected to accommodate quantum mechanics with the hypothesis of electromagnetism.

1948 Solvay Gathering: Post-The Second Great War, this meeting dug into the numerical and reasonable difficulties of bringing together quantum mechanics with the hypothesis of fields. Quantum field hypothesis, which supports QED, got critical consideration.

V. The Impact of Solvay Meetings on Quantum Mechanics

1. **Headway of Quantum Mechanics:**
   The Solvay Gatherings gave a novel stage to driving physicists to trade thoughts and take part in conversations basic to the improvement of quantum mechanics.
   Max Planck, Albert Einstein, Niels Bohr, Werner Heisenberg, Erwin Schrödinger, and different lights effectively partook in these meetings, adding to the refinement of quantum ideas.

2. **Goal of Logical Conflicts:**
   The Solvay Meetings were scenes for logical discussion and the goal of conflicts. The energetic discussions among Einstein and Bohr at the 1927 meeting explained the standards of quantum mechanics and the idea of reality at the quantum level.

3. **Encouraging Cooperation:**
   The Solvay Meetings united physicists from different foundations and ethnicities, cultivating cooperation and cross-treatment of thoughts. These cooperations prompted new bits of knowledge and leap forwards in different areas of physical science.

4. **Setting Exploration Plans:**
   The Solvay Gatherings frequently set the plan for future

exploration in physical science. They distinguished basic inquiries and directed the course of logical request. For example, the 1954 gathering on rudimentary particles assisted shape the examination with programing that prompted the advancement of the Standard Model of molecule material science.

5. **Advancing Interdisciplinary Methodologies:**

The Solvay Gatherings urged interdisciplinary ways to deal with material science. Participants from various subfields of material science met up to examine normal issues and difficulties, working with the improvement of a comprehensive comprehension of the actual universe.

Chapter 6

Einstein's Thought Experiments

Albert Einstein, quite possibly of the most celebrated and powerful physicist ever, changed how we might interpret the universe through his noteworthy hypotheses of relativity and commitments to quantum mechanics. However, similarly striking were his psychological studies — innovative, calculated practices that permitted him to investigate complex actual ideas, challenge the tried and true way of thinking, and show up at significant bits of knowledge. In this investigation, we dive into the universe of Einstein's psychological tests, revealing insight into how these psychological excursions molded the underpinnings of present day physical science.

1. **Presentation: The Force of Psychological studies**
   Psychological studies, otherwise called Gedankenexperiments, are an instrument utilized by researchers and logicians to investigate and test hypothetical ideas. They include making speculative situations in the brain, liberated from the requirements of actual trial and error. Einstein's psychological tests were a sign of his virtuoso, empowering him to picture and figure out the key standards of the universe.

2. **The Extraordinary Hypothesis of Relativity:**

1. **The Train and the Lightning Bolt:**

   Quite possibly of Einstein's most popular psychological study, directed during his young years, laid the preparation for his hypothesis of unique relativity. He envisioned himself riding on a train moving at the speed of light, taking a gander at a mirror, and considering what he would find in the mirror in the event that he were going at the speed of light. This psychological study drove him to consider the steadiness of the speed of light, a foundation of his hypothesis.

2. **The Twin Catch 22:**

Einstein's twin Catch 22 is another notable psychological test. He imagined two indistinguishable twins, one remaining on Earth while the other leaves on a space venture for a huge portion of the speed of light. As per extraordinary relativity, time expansion happens, and the voyaging twin ages all the more leisurely. This Gedankenexperiment delineated the significant impacts of relative movement on time.

III. The Overall Hypothesis of Relativity:

1. **Falling Lift:**

   Einstein's "lift psychological study" was instrumental in fostering the idea of general relativity. He imagined an individual inside a fixed lift in drop. In this situation, the tenant would feel weightless, as though gravity had vanished. This knowledge prompted the equality standard, which turned into a key part of his hypothesis of attractive energy.

2. **Bended Spacetime and the Twisting of Light:**

In one more psychological study, Einstein envisioned a light emission passing near a gigantic item, like the sun. He understood that the way of the light would bend because of the distorting of spacetime brought about by the monstrous article. This psychological activity gave the premise to his overall hypothesis of relativity, which portrays gravity as the ebb and flow of spacetime.

IV. Quantum Mechanics and the EPR Oddity:

While Einstein is most popular for his commitments to relativity, he likewise made critical introductions to the universe of quantum mechanics, though with a portion of doubt.

1. **Einstein-Podolsky-Rosen (EPR) Catch 22:**
   During the 1930s, Einstein, alongside colleagues Podolsky and Rosen, formed a psychological study that tested the culmination of quantum mechanics. They envisioned two snared particles, isolated by an immense distance, with estimations made on one molecule promptly influencing the other, disregarding the standard of territory. The EPR oddity featured the philosophical pressure between quantum entrapment and the idea of neighborhood authenticity.

2. **Einstein's Investigate of Quantum Mechanics:**

Einstein was broadly reproachful of specific parts of quantum mechanics, especially its probabilistic and indeterministic nature. He frequently communicated his uneasiness with the possibility that "God plays dice with the universe." Einstein's protests fixated on the conviction that there should exist stowed away factors that decide the results of quantum occasions, a view he held for the rest of his life.

V. The Podolsky-Rosen-Bohm (EPRB) Trial:

Einstein's study of quantum mechanics and his promotion for buried factors prompted additionally psychological tests, coming full circle in the EPRB try. This theoretical situation included trapped particles and estimations at different points to test whether stowed away factors could make sense of quantum relationships.

VI. Trial Acknowledgment:

The EPRB analyze was subsequently acted during the 1960s by physicist John Chime, who planned Ringer's hypothesis. Ringer's hypothesis gave a structure to test the forecasts of quantum mechanics and secret variable speculations. Tests propelled by Chime's hypothesis reliably upheld the quantum mechanical expectations, testing Einstein's idea of stowed away factors and neighborhood authenticity.

VII. Tradition of Einstein's Psychological tests:

1. **Headway of Key Speculations:**
   Einstein's psychological studies were instrumental in the advancement of both unique and general relativity. These hypotheses reformed how we might interpret space, time, and gravity, prompting expectations that were subsequently affirmed by exploratory proof, like the bowing of light by gravity.

2. **The EPR Oddity and Ringer's Hypothesis:**
   Einstein's psychological studies, especially the EPR oddity, ignited critical interest in the underpinnings of quantum mechanics. The resulting improvement of Ringer's hypothesis and trial of its expectations developed how we might interpret quantum trap and the impediments of neighborhood authenticity.

3. **Effect on Philosophical Discussions:**
   Einstein's psychological tests keep on being at the focal point of philosophical discussions about the idea of the real world and the job of determinism in physical science. His incredulity toward quantum mechanics has incited continuous conversations about the translation of quantum peculiarities.

4. **Instructive Device:**

Einstein's psychological tests act as strong instructive devices, helping understudies and researchers the same accept complex actual ideas. They represent the significance of creative mind and imagination in logical request.

6.1In-depth examination of Einstein's famous thought experiments.

Albert Einstein, perhaps of the most splendid and notable figure throughout the entire existence of science, reshaped how we might interpret the universe through his earth shattering psychological tests. These creative activities permitted him to investigate complex actual ideas, challenge the customary way of thinking, and determine significant experiences that upset the groundworks of physical science. In this top to bottom assessment, we will dig into a portion of Einstein's most popular psychological

studies, analyzing the hidden standards, suggestions, and enduring commitments to our logical comprehension.

1. **Presentation: The Force of Psychological tests**

   Psychological tests, or Gedankenexperiments, are mental activities utilized by researchers and thinkers to examine perplexing or conceptual thoughts without the requirement for actual trial and error. Einstein's psychological studies were a foundation of his logical methodology, showing the significance of creative mind and calculated thinking chasing information.

2. **The Extraordinary Hypothesis of Relativity:**

1. **The Train and the Lightning Bolt:**

   Einstein's excursion into the domain of extraordinary relativity started with a basic yet significant psychological test including a train and a lightning bolt. This trial is a central idea in his hypothesis and enlightens the standards of relativity.

   In this situation, Einstein imagined an onlooker inside a train moving at a steady speed, taking a gander at a mirror situated halfway between two lightning strikes that hit the train tracks at the same time. The psychological study prompted a significant understanding: no matter what the train's movement, the eyewitness inside the train would continuously see the light from both lightning strikes arrive at the mirror simultaneously.

   Suggestion: This psychological study uncovered the consistency of the speed of light, which is an essential propose of unique relativity. Einstein's decision tested traditional thoughts of time and concurrence, making way for his hypothesis of relativity.

2. **The Twin Conundrum:**

Einstein's twin Catch 22 is another famous psychological test that exhibits the impacts of relative movement on time. In this situation, he envisioned two indistinguishable twins: one remaining parts on Earth while the other sets out on a space venture for a critical portion of the speed of light.

As the venturing out twin re-visitations of Earth, he has matured not as much as his Earth-bound kin, in spite of both having encountered the progression of time. This mystery features the peculiarity of time enlargement, a vital standard of extraordinary relativity.

Suggestion: Einstein's twin oddity psychological test represents the overall idea of time and difficulties the old style thought of outright time. It shows that time is certainly not a

general steady yet relies upon the spectator's relative movement, laying the basis for the hypothesis of exceptional relativity.

III. The Overall Hypothesis of Relativity:

1. **The Falling Lift:**

   Einstein's "lift psychological study" was instrumental in fostering the idea of general relativity, which depicts gravity as the bend of spacetime. In this situation, he envisioned an onlooker inside a fixed lift in drop, feeling weightless and encountering no gravitational impacts.

   Suggestion: Einstein's psychological test prompted the proportionality standard, which expresses that locally, it is basically impossible to recognize gravitational powers and speed increase. This rule turned into a central part of his hypothesis of attraction, proposing that monstrous items twist the texture of spacetime, causing gravitational fascination.

2. **Bended Spacetime and the Twisting of Light:**

Einstein's reasonable jump in the improvement of general relativity included envisioning a light emission passing close to a gigantic item, like the sun. He imagined that the way of light would bend because of the distorting of spacetime brought about by the enormous item.

Suggestion: This psychological test delineated that gravity isn't a power in the customary sense yet rather the consequence of items following bended ways in bended spacetime. It anticipated the bowing of starlight by the sun during a sun oriented obscure, a peculiarity affirmed by the 1919 overshadowing endeavor drove by Sir Arthur Eddington, giving exploratory proof to general relativity.

IV. Quantum Mechanics and the EPR Conundrum:

While Einstein is basically connected with the hypothesis of relativity, he likewise wandered into the universe of quantum mechanics, but with a portion of distrust.

1. **Einstein-Podolsky-Rosen (EPR) Mystery:**

   During the 1930s, Einstein, alongside partners Boris Podolsky and Nathan Rosen, figured out a psychological test that tested the culmination of quantum mechanics. They envisioned two caught particles, isolated by a huge distance, with estimations made on one molecule quickly influencing the other, disregarding the rule of territory.

   Suggestion: The EPR mystery featured the philosophical pressure between quantum trap and the idea of neighborhood authenticity. While Einstein and his partners

   contended that quantum mechanics was inadequate, it incited conversations and investigations that investigated the idea of trap and the limits of traditional authenticity.

2. **Einstein's Scrutinize of Quantum Mechanics:**

Einstein was broadly condemning of specific parts of quantum mechanics, especially its probabilistic and indeterministic nature. He frequently communicated his distress with the possibility that "God plays dice with the universe" and accepted that a total hypothesis ought to consider deterministic expectations.

Suggestion: Einstein's investigate of quantum mechanics prodded continuous conversations about the translation of quantum peculiarities, the job of stowed away factors, and the idea of actual reality at the quantum level. While his perspectives didn't line up with the predominant Copenhagen understanding, they added to the philosophical talk encompassing quantum hypothesis.

V. The Podolsky-Rosen-Bohm (EPRB) Examination:

Einstein's evaluate of quantum mechanics and his promotion for buried factors prompted additionally psychological tests, coming full circle in the EPRB analyze. This theoretical situation included trapped particles and estimations at different points to test whether stowed away factors could make sense of quantum connections.

Suggestion: The EPRB psychological test laid the foundation for John Chime's hypothesis, which gave a numerical structure to test the expectations of quantum mechanics and secret variable speculations. Resulting tests enlivened by Ringer's hypothesis reliably upheld the quantum mechanical forecasts, testing Einstein's thought of stowed away factors and nearby authenticity.

VI. Tradition of Einstein's Psychological studies:

1. **Progression of Crucial Hypotheses:**
   Einstein's psychological studies were instrumental in the advancement of both unique and general relativity. These speculations reformed how we might interpret space, time, and gravity, prompting expectations that were subsequently affirmed by exploratory proof.

2. **The EPR Oddity and Chime's Hypothesis:**
   Einstein's psychological studies, especially the EPR Catch 22, started huge interest in the groundworks of quantum mechanics. The resulting advancement of Chime's hypothesis and explor-

atory trial of its forecasts developed how we might interpret quantum entrapment and the constraints of nearby authenticity.

3. **Impact on Philosophical Discussions:**
Einstein's psychological studies keep on being at the focal point of philosophical discussions about the idea of the real world and the job of determinism in physical science. His suspicion toward quantum mechanics has provoked continuous conversations about the translation of quantum peculiarities.

4. **Instructive Apparatus:**

Einstein's psychological tests act as strong instructive apparatuses, helping understudies and researchers the same understand complex actual ideas. They represent the significance of creative mind and imagination in logical request.

6.2 The EPR paradox and its challenge to quantum entanglement.

The Einstein-Podolsky-Rosen (EPR) mystery is a psychological test that arose during the 1930s and represented an essential test to the comprehension of quantum mechanics and the idea of quantum trap. Albert Einstein, Boris Podolsky, and Nathan Rosen figured out this mystery to scrutinize the fulfillment of quantum hypothesis and its suggestions for the idea of actual reality. In this investigation, we will plunge into the EPR Catch 22, its basic standards, and its significant effect on the continuous discussion encompassing quantum ensnarement.

1. **Presentation: The Introduction of the EPR Mystery**
In 1935, Einstein, Podolsky, and Rosen distributed a paper named "Can Quantum-Mechanical Depiction of Actual The truth be Viewed as Complete?" This paper presented the EPR mystery, a psychological study intended to challenge key parts of quantum mechanics, especially the peculiarity of quantum ensnarement.

2. **Quantum Ensnarement: A Preliminary**
Prior to digging into the EPR conundrum, understanding the

idea of quantum entanglement is significant. In quantum mechanics, particles can become entrapped, meaning their properties become connected so that the estimation of one molecule immediately impacts the properties of another, no matter what the distance isolating them. This peculiarity is broadly depicted by the proverb "creepy activity a ways off," an expression begat by Einstein to communicate his disquiet with the ramifications of snare.

3. **The EPR Conundrum Psychological study:**

Einstein, Podolsky, and Rosen planned a psychological test that expected to challenge the fulfillment and ramifications of quantum mechanics. The situation included two ensnared particles, like electrons, ready such that their properties, explicitly their momenta and positions, were related. They contended that the quantum mechanical portrayal of these particles prompted a mystery.

1. **Molecule Relationship:**
   In the EPR try, two snared particles (An and B) are discharged from a typical source and move in inverse headings. Since they are ensnared, they are portrayed by a joint quantum express that mirrors their corresponded properties.

2. **Estimation Proposition:**
   That's what EPR recommended assuming an onlooker estimated the place of molecule A with high accuracy, they could derive the place of molecule B with equivalent accuracy, regardless of whether molecule B was situated far away. Likewise, on the off chance that the eyewitness estimated the force of molecule A, they could decide the energy of molecule B with high accuracy, even without straightforwardly estimating molecule B.

3. **Suggestion: Non-Nearby Associations:**

The EPR conundrum tested the guideline of neighborhood authenticity, which proposes that actual peculiarities are represented by nearby causes and that far off occasions can't impact each other momentarily. Assuming the EPR situation were right, it would infer the presence of non-nearby associations between particles, where the estimation of one molecule quickly decides the condition of the other, no matter what the spatial partition.

IV. The Test to Quantum Mechanics:

The EPR oddity introduced an immediate test to the culmination of quantum mechanics, as it definitely implied that the hypothesis expected extra secret factors or components to represent the relationships between's snared particles.

1. **Einstein's Confidence in Determinism:**
   Einstein was a firm promoter of determinism, the possibility that the result of any actual not set in stone by previous circumstances. He trusted that quantum mechanics, with its innate probabilistic nature, was a fragmented portrayal of actual reality. He broadly jested, "God doesn't play dice with the universe," communicating his distress with the arbitrariness intrinsic in quantum hypothesis.

2. **Secret Factors Theory:**

Einstein, Podolsky, and Rosen proposed the presence of stowed away factors — undetectable properties of particles that decide their way of behaving and represent the relationships saw in ensnared frameworks. That's what they contended in the event that these secret factors were known, the clear non-nearby associations between particles could be made sense of in a deterministic, neighborhood way.

V. Reaction from the Quantum People group:

The EPR mystery touched off serious discussions inside the material science local area and set off a reaction from quantum scholars, most strikingly Niels Bohr and Werner Heisenberg, who were defenders of the Copenhagen translation of quantum mechanics.

1. **Bohr's Reaction:**

   Niels Bohr, a focal figure in the improvement of quantum mechanics, answered the EPR conundrum by safeguarding the standards of complementarity and the Copenhagen understanding. He contended that the EPR situation didn't challenge the culmination of quantum mechanics yet rather featured the restrictions of old style ideas in portraying quantum peculiarities.

2. **Complementarity:**

Bohr presented the idea of complementarity, which recommends that specific sets of reciprocal properties, like position and energy, can't be at the same time estimated with inconsistent accuracy. He fought that the demonstration of estimation itself impacted the properties being noticed, and traditional ideas of determinism and region didn't make a difference at the quantum level.

VI. Chime's Hypothesis:

The EPR conundrum kept on charming physicists, prompting further examinations concerning the idea of quantum snare. During the 1960s, physicist John Chime formed a hypothesis that offered an approach to test the forecasts of quantum mechanics and secret variable speculations tentatively.

1. **Chime's Imbalance:**

   Ringer's hypothesis gave a numerical system to test whether the connections saw in entrapped particles could be made sense of by stowed away factors. He determined a bunch of imbalances, known as Chime disparities, that should turn out as expected assuming that the relationships between's entrapped particles were made sense of by nearby secret factors.

2. **Trial Tests:**

Tests enlivened by Chime's hypothesis reliably abused the Ringer imbalances, showing that the connections between's caught particles couldn't be made sense of by neighborhood stowed away factors. These investigations offered solid exact help for quantum entrapment and disproved the EPR proposition for buried factors.

VII. The Angle Investigation and Infringement of Ringer's Disparities:

In 1982, physicist Alain Perspective directed a weighty trial known as the Viewpoint explore, which tried Ringer's imbalances. The aftereffects of the examination indisputably exhibited that the forecasts of quantum mechanics were contradictory with neighborhood authenticity. Entrapped particles displayed relationships that surpassed the cutoff points set by Chime's imbalances, affirming the non-neighborhood nature of quantum snare.

VIII. Suggestions and Translations:

The goal of the EPR oddity and the infringement of Ringer's disparities have significant ramifications for how we might interpret quantum mechanics and the idea of actual reality:

1. **Non-Territory and Snare:**
   The EPR oddity and resulting tests have laid out that quantum trap includes non-nearby relationships between's particles, testing traditional ideas of territory and determinism. This non-territory is a crucial element of the quantum world.

2. **Quantum Mechanics as a Total Hypothesis:**
   The infringement of Ringer's imbalances gave undeniable proof that quantum mechanics is a finished and self-steady hypothesis that precisely depicts the way of behaving of particles at the quantum level. It doesn't need stowed away factors to make sense of the noticed peculiarities.

3. **Interpretational Discussions:**

While the EPR mystery has been settled for quantum mechanics, interpretational banters about the idea of quantum reality persevere. Different understandings, like the Many-Universes translation, Bohmian mechanics, and the goal breakdown models, keep on investigating the philosophical ramifications of quantum ensnarement.

6.3 Einstein's attempts to demonstrate the incompleteness of quantum mechanics.

Albert Einstein, prestigious for his commitments to the hypothesis of relativity, was likewise a vital figure in the early improvement of quantum mechanics. While quantum mechanics has turned into a fundamental hypothesis in material science, Einstein held second thoughts about its culmination. All through his profession, he participated in different psychological tests and discussions, endeavoring to exhibit the apparent deficiency of quantum mechanics. In this investigation, we will dive into Einstein's endeavors to challenge the fulfillment of quantum mechanics and the philosophical discussions that resulted.

1. **Presentation: Einstein and Quantum Mechanics**

   Einstein's contribution with quantum mechanics started with his noteworthy work on the photoelectric impact in 1905, which gave essential proof to the quantum idea of light. Notwithstanding, as quantum mechanics advanced, Einstein turned out to be progressively suspicious of specific parts of the hypothesis.

2. **Einstein-Podolsky-Rosen (EPR) Oddity:**

The EPR mystery, presented by Einstein, Boris Podolsky, and Nathan Rosen in their 1935 paper, "Can Quantum-Mechanical Depiction of Actual The truth be Thought of as Complete?," is maybe the most popular illustration of Einstein's endeavors to challenge quantum mechanics.

1. **The EPR Catch 22 Situation:**

   In the EPR Catch 22, the writers thought about two particles (e.g., electrons) that were ready in a snared state. This implies that their properties, like twist, were connected, in any event, when isolated by huge distances.

2. **EPR's Investigate: Non-Territory and Secret Factors:**
Einstein, Podolsky, and Rosen contended that quantum mechanics anticipated a degree of connection between's these particles that couldn't be made sense of by neighborhood stowed away factors — inconspicuous properties of particles that decide their way of behaving. They suggested that such secret factors ought to exist to give a more deterministic clarification to these connections.

3. **Suggestion: The Test to Quantum Mechanics:**

The EPR mystery tested the culmination of quantum mechanics, recommending that there may be covered up factors or factors not represented in the hypothesis. In the event that such secret factors existed, they might actually give a more complete

portrayal of actual reality, as opposed to the probabilistic and indeterministic nature of quantum mechanics.

III. Bohr's Reaction:

Niels Bohr, a main figure in quantum mechanics and defender of the Copenhagen translation, participated in a progression of discussions with Einstein over the EPR conundrum. Bohr's reaction was critical in forming the talk on quantum mechanics and its culmination.

1. **Complementarity and the Copenhagen Translation:**
Bohr protected the Copenhagen understanding, which held that quantum mechanics gave a total and self-reliable depiction of actual reality, but one that frequently challenged traditional instinct. He presented the idea of complementarity, proposing that quantum peculiarities could show both wave-like and molecule like properties, contingent upon the trial arrangement.

2. **Bohr's Contention: Non-Area and Ensnarement:**

Bohr battled that the EPR mystery didn't challenge quantum mechanics' fulfillment yet rather exhibited the restrictions of old style ideas in portraying quantum peculiarities. He contended that non-neighborhood relationships, as seen in snared frameworks, were a central element of quantum mechanics and were not in struggle with its culmination.

IV. Ringer's Hypothesis:

The discussion among Einstein and Bohr over the EPR mystery laid the foundation for additional examinations concerning the idea of quantum entrapment. Physicist John Ringer figured out a hypothesis during the 1960s that gave a structure to trial tests to decide if quantum mechanics or secret factors could make sense of the noticed connections in caught frameworks.

1. **Ringer's Disparity:**

   Ringer's hypothesis laid out a bunch of disparities, known as Chime imbalances, that should turn out as expected on the off chance that the relationships saw in trapped particles could be made sense of by nearby secret factors.

2. **Exploratory Tests:**

Tests enlivened by Chime's hypothesis reliably abused these disparities, showing the way that quantum mechanics couldn't be made sense of by nearby secret factors. These outcomes offered solid observational help for the non-neighborhood nature of quantum snare.

V. Einstein's Proceeding with Scrutinize:

Indeed, even despite Chime's hypothesis and exploratory proof supporting quantum mechanics, Einstein kept on evaluating specific parts of the hypothesis. He had some lingering doubts of the probabilistic and indeterministic nature of quantum mechanics and tried to investigate elective clarifications.

1. **"God Doesn't Play Dice":**
   Einstein broadly pronounced, "God doesn't play dice with the universe," communicating his distress with the intrinsic arbitrariness in quantum mechanics. He kept up with his confidence in the presence of stowed away factors that could decide the results of quantum occasions in a deterministic way.
2. **Einstein's Joint effort with De Broglie:**

In the last part of the 1940s, Einstein teamed up with individual physicist Louis de Broglie trying to foster a hypothesis that could accommodate quantum mechanics with deterministic secret factors. Their endeavors prompted the improvement of the "Einstein-de Broglie pilot-wave hypothesis," otherwise called "Bohmian mechanics."

VI. Bohmian Mechanics:

Bohmian mechanics, created during the 1950s by David Bohm, was enlivened by crafted by Einstein and de Broglie. This hypothesis presented a deterministic system for quantum mechanics, including stowed away factors and pilot waves that directed the way of behaving of particles.

1. **Deterministic Directions:**
   In Bohmian mechanics, particles have distinct positions and momenta, with not set in stone by a directing wave capability. This system planned to give a deterministic clarification to quantum peculiarities while safeguarding the expectations of quantum mechanics.
2. **Restricted Acknowledgment:**

Bohmian mechanics, while charming, didn't acquire far and wide acknowledgment among the material science local area. Numerous physicists inclined toward the Copenhagen understanding and the probabilistic idea of quantum mechanics, seeing Bohmian mechanics as a superfluous takeoff from laid out hypothesis.

7

Chapter 7

Bohr's Response

Niels Bohr and Albert Einstein were two transcending figures in the improvement of quantum mechanics during the mid twentieth 100 years. While both made huge commitments to the field, they had differentiating perspectives on the idea of quantum reality. Einstein, known for his doubt about the fulfillment of quantum mechanics, participated in a few discussions with Bohr, a defender of the Copenhagen translation. In this investigation, we will dive into Bohr's reaction to Einstein's difficulties and the guard of the Copenhagen translation, a reaction that keeps on forming how we might interpret the quantum world.

1. **Presentation: The Quantum Upset and Einstein's Distrust**
   The mid twentieth century denoted a time of significant disturbance in the realm of physical science, with the development of quantum mechanics testing old style physical science and our essential comprehension of the universe. Quantum mechanics presented the possibility that particles at the quantum level don't have distinct properties until estimated, leading to the popular expression "breakdown of the wave capability." Nonetheless, this calculated shift didn't agree with Einstein, who broadly

communicated his inconvenience with the probabilistic and indeterministic nature of quantum mechanics.

2. The Copenhagen Understanding of Quantum Mechanics:

At the very front of the improvement of quantum mechanics was the Copenhagen translation, named after the city where Niels Bohr and Werner Heisenberg, among others, established its groundworks. The Copenhagen translation is a structure for figuring out quantum peculiarities and the philosophical ramifications of the hypothesis. Key components of the understanding include:

1. **Wave-Molecule Duality:**
   The Copenhagen translation recognizes the wave-molecule duality of quantum objects. Particles, for example, electrons show both wave-like and molecule like properties, contingent upon the exploratory setting.

2. **Complementarity:**
   Bohr presented the idea of complementarity, which proposes that specific sets of reciprocal properties, like position and force, can't be all the while estimated with inconsistent accuracy. This idea mirrors that quantum peculiarities can be seen according to alternate points of view, each noteworthy just a fractional truth.

3. **Vulnerability Standard:**

Heisenberg's vulnerability rule is a crucial part of the Copenhagen translation. It expresses that the more definitively the place of a molecule is known, the less exactly its force can be known, as well as the other way around. This guideline underlines the innate constraints of estimation in the quantum world.

III. Einstein's Difficulties to Quantum Mechanics:

Albert Einstein, albeit a trailblazer of quantum hypothesis, held profound second thoughts about specific parts of quantum mechanics, especially its probabilistic and indeterministic nature. His essential worries included:

1. **Determinism:**
   Einstein was a defender of determinism, the possibility that the result of any actual not entirely settled by prior conditions. He trusted that quantum mechanics, with its innate haphazardness and eccentricism, was inadequate and ought to at last take into consideration deterministic expectations.

2. **Secret Factors:**

Einstein recommended the presence of stowed away factors — imperceptible properties of particles that decide their way of behaving. He contended that these secret factors could give a more complete and deterministic depiction of quantum peculiarities.

IV. The Einstein-Podolsky-Rosen (EPR) Conundrum:

One of Einstein's most huge difficulties to quantum mechanics was the EPR conundrum, figured out as a team with Boris Podolsky and Nathan Rosen in their 1935 paper, "Can Quantum-Mechanical Depiction of Actual The truth be Thought of as Complete?" The EPR mystery proposed a situation including two entrapped particles and contended that the quantum mechanical portrayal of these particles prompted perplexing results.

1. **The EPR Oddity Situation:**
   In the EPR explore, two entrapped particles are ready such that their properties, like twist, are connected. In any event, when isolated by huge distances, estimations on one molecule momentarily influence the properties of the other, abusing the guideline of area.

2. **EPR's Scrutinize: Non-Region and Secret Factors:**
   Einstein, Podolsky, and Rosen contended that the EPR situation tested the fulfillment of quantum mechanics, recommending

that secret factors could exist to give a more deterministic clarification to these relationships. They suggested that such secret factors ought to exist to save the guideline of area.

3. **Suggestion: The Test to Quantum Mechanics:**

The EPR mystery tested the culmination of quantum mechanics, as it implied that the hypothesis expected extra secret factors or components to represent the connections between's snared particles.

V. Bohr's Reaction to the EPR Oddity:

Niels Bohr, one of the main defenders of the Copenhagen understanding, took part in a progression of discussions with Einstein over the EPR mystery and the more extensive philosophical ramifications of quantum mechanics. Bohr's reaction was crucial in molding the talk on quantum mechanics and its fulfillment.

1. **Complementarity and the Copenhagen Understanding:**
   Bohr safeguarded the Copenhagen translation as a total and self-predictable depiction of actual reality, even despite the EPR oddity. He contended that the conundrum didn't challenge quantum mechanics' culmination yet rather showed the constraints of old style ideas in depicting quantum peculiarities.

2. **Complementarity:**
   Bohr presented the idea of complementarity as a key element of quantum mechanics. Complementarity recommends that specific sets of integral properties, like position and force, can't be at the same time estimated with erratic accuracy. The demonstration of estimation itself impacts the properties being noticed.

3. **Non-Region and Ensnarement:**
   Bohr battled that the EPR mystery didn't challenge quantum mechanics' culmination yet rather exhibited the non-nearby nature of quantum trap, where estimations on one molecule could promptly impact the properties of another, no matter what the spatial

division. He contended that the idea of region was generally unique at the quantum level.

4. The Spectator Impact:

Bohr stressed that the demonstration of estimation was a basic piece of quantum mechanics and couldn't be isolated from the peculiarities being noticed. He kept up with that quantum mechanics gave a total depiction of the real world, however this the truth was innately not the same as old style instincts.

VI. Bohr's Impact and Inheritance:

Niels Bohr's protection of the Copenhagen understanding and his reactions to Einstein's difficulties impacted the improvement of quantum mechanics and its translation:

1. **Molding the Copenhagen Translation:**
   Bohr's enunciation of the Copenhagen translation as a total and self-predictable system for understanding quantum peculiarities hardened its position in the records of physical science. It stays one of the most generally instructed and applied understandings of quantum mechanics.

2. **Philosophical Effect:**

Bohr's accentuation on complementarity, the non-nearby nature of quantum trap, and the connection of the eyewitness from the noticed significantly affect the way of thinking of science. These ideas keep on molding conversations about the idea of reality at the quantum level.

7.1Bohr's responses to Einstein's thought experiments and critiques.

The discussion between Albert Einstein and Niels Bohr over the underpinnings of quantum mechanics remains as perhaps of the most huge and persevering through conversation throughout the entire existence of material science. Einstein, known for his wariness about the fulfillment of quantum mechanics, took part in a progression of psychological tests and evaluates pointed toward testing the overall understanding of quantum hypothesis. On the opposite side, Niels Bohr, a vital defender of the Copenhagen translation, answered Einstein's difficulties with a blend of philosophical meticulousness and protection of the probabilistic and indeterministic nature of quantum mechanics. In this investigation, we will dive into Bohr's reactions to Einstein's psychological studies and scrutinizes, revealing insight into the complexities of their scholarly conflict.

1. **Presentation: The Conflict of Titans**

   The conflict among Einstein and Bohr was established in their varying philosophical positions on the idea of reality in the quantum world. Einstein was profoundly awkward with the probabilistic and indeterministic parts of quantum mechanics, while Bohr guarded the Copenhagen understanding, which underscored the constraints of old style ideas in depicting quantum peculiarities.

2. **Einstein's Psychological tests and Evaluates:**

Einstein's evaluates of quantum mechanics frequently appeared as psychological studies, intended to feature what he saw as lacks in the hypothesis. These psychological tests covered a scope of subjects, from determinism to the idea of particles. The absolute most eminent ones include:

1. **The Photon Box Analysis:**

   In a 1905 psychological study, Einstein imagined a fixed box with entirely reflecting walls in which a photon of light was bobbing between the walls. That's what he contended assuming the photon's energy relied upon its recurrence, as recommended by Max Planck's quantum hypothesis, then the energy inside the case would be questionable. This psychological study alluded to the possibility that quantum mechanics brought innate vulnerabilities into actual frameworks.

2. **The EPR Conundrum:**

The Einstein-Podolsky-Rosen (EPR) conundrum, formed in 1935, tested quantum mechanics' culmination. In this psychological test, two entrapped particles, isolated by a significant stretch, display corresponded properties. Einstein, alongside Boris Podolsky and Nathan Rosen, contended that this inferred the presence of stowed away factors, properties that decided the particles' way of behaving and could make sense of their connections without disregarding area.

3. **The Photon and Gravitational Redshift:**

In one more psychological test, Einstein considered a photon going vertical and away from a monstrous body like Earth. That's what he contended, as per quantum mechanics, the photon's energy would be quantized and, thus, it would redshift because of its departure from the gravitational field. This, he accepted, prompted a conundrum since it went against the standard of equality in his hypothesis of general relativity.

III. Bohr's Reactions:

Niels Bohr answered Einstein's psychological tests and evaluates with a mix of philosophical thinking and protection of the Copenhagen understanding. His reactions frequently underscored the non-traditional nature of quantum peculiarities and the restrictions of old style ideas in portraying them:

1. **Complementarity:**

Bohr presented the idea of complementarity, which recommended that specific sets of correlative properties, like position and force, couldn't be all the while estimated with erratic accuracy. This idea underlined that quantum peculiarities could show both wave-like and molecule like properties, contingent upon the trial setting. Complementarity was Bohr's reaction to Einstein's photon box explore, as it focused on the need to see quantum peculiarities according to alternate points of view.

2. **The EPR Conundrum and Non-Area:**

Bohr answered the EPR conundrum by underlining the non-nearby nature of quantum entrapment. He contended that the relationships saw in entrapped frameworks were not in struggle with quantum mechanics' culmination but rather showed the restrictions of old style ideas like area. As indicated by Bohr, the idea of non-region was essentially unique at the quantum level, and quantum mechanics gave a total depiction of this reality.

3. **The Photon and Gravitational Redshift:**

Because of Einstein's psychological study on the gravitational redshift of a photon, Bohr stressed the comparability guideline, an essential idea in Einstein's hypothesis of general relativity. Bohr contended that the identicalness guideline held in quantum mechanics and that the photon's energy quantization and redshift were reliable with it. He focused on that traditional and quantum portrayals of actual peculiarities were intrinsically unique, and the idea of quantization was fundamental in the quantum domain.

IV. The Bohr-Einstein Discussions:

The discussions among Bohr and Einstein were logical conversations as well as philosophical conflicts. Their conflicts addressed significant inquiries regarding the idea of the real world, determinism, and the job of eyewitnesses in quantum mechanics.

1. **Bohr's Philosophical Position:**

Bohr's reactions frequently highlighted the significance of embracing the probabilistic and indeterministic parts of quantum mechanics. He contended that old style thoughts of determinism and causality were not pertinent at the quantum level and that the demonstration of estimation was a basic piece of quantum peculiarities.

2. Einstein's Journey for Determinism:

Einstein, conversely, was persuaded by a firmly established faith in determinism, the possibility that actual occasions had unequivocal, previous causes. He was awkward with the probabilistic idea of quantum mechanics and looked for buried factors or deterministic clarifications that would line up with his philosophical convictions.

V. Bohr's Impact and Inheritance:

Niels Bohr's reactions to Einstein's scrutinizes set the Copenhagen translation as a primary system for figuring out quantum mechanics. His accentuation on complementarity, non-territory, and the spectator impact molded the manner in which physicists moved toward the quantum world.

1. **Impact on Translations of Quantum Mechanics:**
   The Copenhagen understanding, with its acknowledgment of probabilistic results and the job of the onlooker, stays one of the most generally educated and applied translations of quantum mechanics. Bohr's accentuation on non-region likewise laid the foundation for additional investigation of ensnarement and Ringer's hypothesis.

2. **Proceeding with Philosophical Conversations:**

While Bohr's reactions laid out the Copenhagen understanding, they didn't completely determine the philosophical discussions encompassing quantum mechanics. Inquiries concerning the idea of quantum reality, the job of stowed away factors, and the connection among old style and quantum ideas keep on being subjects of continuous philosophical investigation.

7.2His defense of the Copenhagen interpretation and quantum entanglement.

Niels Bohr, one of the principal architects of quantum mechanics, assumed a focal part in molding the Copenhagen translation, a crucial structure for grasping the quantum world. He protected this understanding resolutely, and his commitments were especially huge in making sense of the strange peculiarity of quantum entrapment. In this investigation, we will dive into Niels Bohr's safeguard of the Copenhagen understanding and his significant bits of knowledge into the idea of quantum ensnarement.

1. **Presentation: The Requirement for Translation**

   Quantum mechanics, with its probabilistic nature and wave-molecule duality, achieved a significant upheaval in material science during the mid twentieth hundred years. Notwithstanding, it likewise presented a serious level of reasonable intricacy and weirdness, provoking the requirement for translation. The Copenhagen translation, created by Bohr and Werner Heisenberg, expected to give an intelligent structure to figuring out quantum peculiarities.

2. **The Copenhagen Understanding: An Outline**

1. **Complementarity:**

   One of the focal principles of the Copenhagen understanding is complementarity, presented by Bohr. It states that specific sets of correlative properties, like position and energy, can't be all the while estimated with inconsistent accuracy. Complementarity mirrors that quantum peculiarities can be seen according to alternate points of view, each noteworthy just an incomplete truth.

2. **Wave-Molecule Duality**

   The Copenhagen understanding recognizes the wave-molecule duality of quantum objects. Particles like electrons can show both wave-like and molecule like properties, contingent upon the exploratory setting. Bohr accentuated that quantum elements didn't have obvious properties until they were estimated.

3. **The Onlooker Impact:**

Bohr focused on the meaning of the eyewitness in quantum mechanics. He contended that the demonstration of estimation affected the properties being noticed and that quantum peculiarities couldn't be completely perceived disregarding the eyewitness' job.

III. Bohr's Safeguard of the Copenhagen Understanding:

Niels Bohr energetically safeguarded the Copenhagen translation against different difficulties and reactions, especially those presented by Albert Einstein. His reactions featured the crucial contrasts between the old style and quantum universes and stressed the need to embrace the probabilistic and indeterministic parts of quantum mechanics.

1. **Embracing Complementarity:**
   Bohr's guard of complementarity was a foundation of the Copenhagen understanding. He contended that specific sets of correlative properties, like position and energy, were innately contrary when estimated with high accuracy. This idea highlighted that quantum

   peculiarities could display inconsistent ways of behaving relying upon the picked trial arrangement.

2. **Non-Traditional Nature:**
   Bohr focused on that old style ideas and instincts couldn't be straightforwardly applied to the quantum world. He accepted that quantum mechanics was a finished and self-reliable hypothesis, however it expected a change in thinking to get a handle on its suggestions completely. Bohr's safeguard of the non-traditional nature of quantum reality filled in as a basic reaction to Einstein's emphasis on deterministic secret factors.

3. **Non-Territory and Entrapment:**

Maybe one of Bohr's most critical commitments was his reaction to quantum entrapment. Ensnarement is a peculiarity where at least two particles become connected so that estimating one molecule in a flash influences the properties of the others, no matter what the distance isolating them. Bohr embraced ensnarement as a central component of quantum mechanics and underscored its non-nearby nature.

IV. Quantum Entrapment: A More critical Look

Bohr's safeguard of quantum entrapment was a critical part of his reaction to Einstein's investigates. Quantum entrapment is a peculiarity that tested traditional instincts about the distinguishableness and freedom of far off objects. Central issues with all due respect of trap include:

1. **Non-Region:**
   Bohr acknowledged the non-area intrinsic in quantum ensnarement, where estimations on one molecule could promptly impact the properties of another, no matter what the spatial partition. He contended that this non-region was an essential part of the quantum world, one that withdrew from old style thoughts of territory and causality.

2. **Indivisibility:**
   Bohr focused on that entrapped particles were intrinsically interconnected, and their properties were not resolved separately however in relationship with one another. This indistinguishability tested traditional thoughts of unmistakable, free elements and highlighted the non-old style nature of quantum peculiarities.

3. **The EPR Catch 22:**

Bohr's reaction to the Einstein-Podolsky-Rosen (EPR) oddity, which presented difficulties to quantum mechanics' culmination, focused on quantum ensnarement. He

contended that the EPR Catch 22 didn't challenge the hypothesis' culmination yet rather exhibited the limits of old style ideas in portraying quantum connections.

V. Examinations and Checks:

Bohr's guard of quantum snare was subsequently upheld by trial proof. The infringement of Chime's imbalances in tests enlivened by

John Ringer's hypothesis affirmed that quantum ensnarement showed non-neighborhood relationships that old style material science couldn't make sense of. These exploratory checks supported Bohr's position on the non-traditional nature of quantum ensnarement.

VI. Bohr's Philosophical Effect:

Bohr's safeguard of the Copenhagen translation and his experiences into quantum entrapment logically affected the understanding of quantum mechanics and the way of thinking of science in general:

1. **Impact on Translations:**
   Bohr's accentuation on complementarity, non-area, and the spectator impact fundamentally affected translations of quantum mechanics. The Copenhagen understanding, with its acknowledgment of probabilistic results and the job of the eyewitness, stays one of the most generally instructed and applied translations of quantum mechanics. Bohr's accentuation on non-territory additionally laid the foundation for additional investigation of ensnarement and Ringer's hypothesis.

2. **Proceeding with Philosophical Conversations:**

While Bohr's reactions laid out the Copenhagen translation, they didn't completely determine the philosophical discussions encompassing quantum mechanics. Inquiries regarding the idea of quantum reality, the job of stowed away factors, and the connection among old style and quantum ideas keep on being subjects of progressing philosophical investigation.

7.3 The evolution of Bohr's ideas in response to Einstein's challenges.

The conflict between Niels Bohr and Albert Einstein over the understanding of quantum mechanics is quite possibly of the most persevering and significant discussion throughout the entire existence of physical science. While Bohr safeguarded the probabilistic and indeterministic nature of quantum mechanics through the Copenhagen understanding, Einstein still had a few doubts and looked for deterministic clarifications. Throughout their long term scholarly trade, Bohr's thoughts developed in light of Einstein's difficulties. In this investigation, we will dig into the development of Bohr's

thoughts as he answered Einstein's studies and looked to accommodate quantum mechanics with the old style world.

1. **Presentation: The Quantum-Exemplary Gap**

   At the core of the Bohr-Einstein banter was the obvious clash between the odd and illogical universe of quantum mechanics and the old style physical science that had ruled logical idea for a really long time. Bohr, alongside Werner Heisenberg and others, fostered the Copenhagen translation to address the oddities and idiosyncrasies of the quantum world. Einstein, notwithstanding, remained profoundly awkward with the probabilistic and in-deterministic parts of quantum mechanics.

2. **Bohr's Initial Guard of the Copenhagen Understanding:**

In the beginning phases of the discussion, Bohr ardently protected the Copenhagen translation, accentuating its essential standards, including complementarity, wave-molecule duality, and the job of the spectator. His underlying reactions to Einstein's provokes mirrored an enduring obligation to the probabilistic and non-deterministic nature of quantum mechanics:

1. **Complementarity as a Support point:**

   Bohr maintained the guideline of complementarity, which he presented, as a basic part of quantum mechanics. Complementarity contended that specific sets of reciprocal properties, similar to position and force, couldn't be all the while estimated with inconsistent accuracy. Bohr saw this as a key element that featured the impediments of traditional ideas when applied to the quantum domain.

2. **Wave-Molecule Duality:**
   Bohr reaffirmed the wave-molecule duality of quantum objects, underscoring that particles, for example, electrons showed both wave-like and molecule like properties, contingent upon the exploratory setting. He kept up with that these dualities were characteristic for the quantum world and couldn't be made sense of utilizing old style ideas.

3. **The Job of the Onlooker:**

Bohr focused on that the demonstration of estimation was a fundamental piece of quantum mechanics. He contended that the spectator's contribution affected the properties being noticed and that traditional ideas of determinism and causality didn't have any significant bearing in the quantum domain. This position was integral to Bohr's protection of the Copenhagen translation.

III. The Einstein-Podolsky-Rosen (EPR) Catch 22:
One of the crucial minutes in the Bohr-Einstein banter was the detailing of the EPR oddity in 1935. In their paper, Einstein, Boris Podolsky, and Nathan Rosen contended that quantum mechanics couldn't be viewed as a total hypothesis since it considered non-nearby relationships between's caught particles. Bohr's reaction to the EPR conundrum denoted a critical development in his thoughts:

1. **Non-Area and Secret Factors:**
   The EPR oddity tested the fulfillment of quantum mechanics by recommending the presence of stowed away factors — undetectable properties of particles that could decide their way of behaving. Bohr's underlying reaction was to underscore that quantum mechanics gave a total and self-steady portrayal of the real world. Notwithstanding, he perceived that the EPR Catch 22 brought up significant issues about non-area.

2. **Embracing Non-Area:**

As the discussion advanced, Bohr's position developed. He came to acknowledge the non-area inferred by quantum snare, where estimations on one molecule could momentarily impact the properties of another, no matter what the spatial partition. Bohr contended that the idea of region was essentially unique at the quantum level, and this acknowledgment denoted a huge change in his reasoning.

IV. Complementarity Returned to:

1. **Complementarity as a Structure:**
   Bohr enunciated complementarity as a structure for figuring out quantum peculiarities, especially those including non-region and ensnarement. He contended that complementarity took into consideration the conjunction of apparently problematic parts of quantum reality, giving a more exhaustive comprehension of the quantum world.

2. **Philosophical Ramifications:**

Bohr's investigation of complementarity had significant philosophical ramifications. He underscored that the impediments of traditional ideas in depicting quantum peculiarities didn't suggest a deficient hypothesis. All things considered, he fought that quantum mechanics uncovered the insufficiency of old style instincts and required a better approach for contemplating the idea of the real world.

V. Bohr's Impact and Heritage:

Niels Bohr's developing thoughts because of Einstein's difficulties lastingly affected the translation of quantum mechanics and the way of thinking of science:

1. **Effect on Translations:**
   Bohr's accentuation on complementarity, non-area, and the acknowledgment of quantum trap altogether impacted translations of quantum mechanics. The Copenhagen translation, with its acknowledgment of probabilistic results and the job of the

spectator, stays one of the most generally instructed and applied understandings of quantum mechanics.

2. **Philosophical Effect:**

Bohr's developing thoughts likewise made a significant imprint on the way of thinking of science. His acknowledgment of the impediments of traditional ideas and the acknowledgment of non-area reshaped conversations about the idea of the real world and the connection among old style and quantum universes.

Chapter 8

Legacy and Impact

The tradition of Niels Bohr's reactions to Albert Einstein's difficulties in the field of quantum mechanics is significant and extensive. The Bohr-Einstein discusses, which traversed quite a few years and resolved major inquiries concerning the idea of the real world and the translation of quantum hypothesis, made a permanent imprint on the field of physical science and the way of thinking of science. In this investigation, we will dive into the heritage and effect of Bohr's reactions to Einstein's difficulties, looking at how these discussions keep on forming how we might interpret the quantum world, impact logical request, and rouse continuous philosophical conversations.

1. **Presentation: The Meaning of the Bohr-Einstein Discussions**
   The Bohr-Einstein discusses were not just learned fighting between two conspicuous physicists; they were a pot where the groundworks of quantum mechanics were tried and refined. At the core of their conflicts were significant inquiries concerning determinism, authenticity, and the idea of actual reality in the quantum domain. The heritage and effect of these discussions are

apparent in different aspects of science, reasoning, and, surprisingly, mainstream society.

2. The Copenhagen Translation as a Foundation:

One of the main traditions of Bohr's reactions to Einstein's difficulties is the getting through impact of the Copenhagen translation of quantum mechanics. Bohr's protection of the probabilistic and indeterministic nature of quantum hypothesis, combined with his hug of ideas like complementarity and non-territory, cemented the Copenhagen translation as a basic system for grasping quantum peculiarities.

1. **Probabilistic Nature of Quantum Mechanics:**
   Bohr's steadfast safeguard of the probabilistic idea of quantum mechanics, where results are portrayed by probabilities instead of deterministically unsurprising, has turned into a key component of present day quantum hypothesis. This probabilistic system supports the reasonable utilizations of quantum mechanics, from quantum processing to quantum cryptography.

2. **Complementarity as a Core value:**
   Bohr's idea of complementarity, which highlights that specific sets of corresponding properties can't be at the same time estimated with inconsistent accuracy, lastingly affects the translation of quantum peculiarities. It gives a system to figuring out the wave-molecule duality of quantum elements and has impacted resulting understandings of quantum mechanics.

3. **Non-Region and Entrapment:**

Bohr's acknowledgment of non-region, as shown by quantum trap, has made ready for the investigation of non-old style relationships and their pragmatic applications. The peculiarity of trap is at the core of quantum advancements, for example, quantum instant transportation and quantum key conveyance, with significant ramifications for secure correspondence and future registering standards.

III. Philosophical Ramifications:

The Bohr-Einstein discusses significantly affected the way of thinking of science, testing conventional thoughts of determinism and authenticity. Bohr's reactions, especially his accentuation on the restrictions of old style ideas and the job of the eyewitness, keep on moving philosophical conversations about the idea of the real world and the connection between the quantum and traditional universes.

1. **Quantum Authenticity versus Hostile to Authenticity:**
   Bohr's safeguard of the Copenhagen translation, with its emphasis on the eyewitness' impact and the deficiency of traditional ideas, led to banters about authenticity versus hostile to authenticity in the way of thinking of science. Logicians keep on investigating whether quantum mechanics gives a genuine portrayal of an objective reality or simply a numerical instrument for anticipating results.

2. **Difficulties to Traditional Instincts:**

Bohr's acknowledgment of the constraints of traditional ideas in depicting quantum peculiarities tested profoundly imbued old style instincts about the idea of the real world. This change in context has provoked continuous philosophical investigations into the connection among traditional and quantum material science and the limits of human getting it.

IV. Effect on Quantum Data Science:

Bohr's reactions to Einstein's difficulties have essentially impacted the field of quantum data science, which investigates the use of quantum mechanics to data handling, calculation, and cryptography. The probabilistic and non-nearby nature of quantum

peculiarities, as guarded by Bohr, shapes the reason for some quantum data advancements.

1. **Quantum Figuring:**
   Bohr's acknowledgment of the probabilistic idea of quantum mechanics established the groundwork for quantum registering, a field that use quantum peculiarities like superposition and snare to perform calculations with possible benefits over old style PCs. Quantum registering can possibly alter fields like cryptography, improvement, and materials science.

2. **Quantum Cryptography:**

Bohr's acknowledgment of the security presented by quantum trap has roused the advancement of quantum cryptography conventions, which use the standards of quantum mechanics to make rugged codes and secure correspondence channels. These advancements can possibly reshape the scene of secure interchanges.

V. Test Approval:

The tradition of Bohr's reactions to Einstein's difficulties isn't restricted to hypothetical discussions; it is additionally obvious in trial checks of the standards he safeguarded. The infringement of Ringer's imbalances in tests roused by John Chime's hypothesis has offered experimental help for the non-neighborhood relationships anticipated by quantum snare, avowing Bohr's position on non-region.

1. **Ringer's Hypothesis and Examinations:**
   Chime's hypothesis, planned during the 1960s, demonstrated the way that no hypothesis in light of nearby secret factors could recreate the relationships anticipated by quantum mechanics. Resulting tests testing Chime's disparities affirmed the non-nearby nature of quantum snare, lining up with Bohr's acknowledgment of non-territory.

2. **Legitimizing Quantum Non-Territory:**

These trial approvals of non-region have supported the Copenhagen translation as well as added to a more profound comprehension of the non-traditional highlights of the quantum world. They have additionally hardened Bohr's place that quantum mechanics gives a precise portrayal of actual reality.

VI. Progressing Discussions and Open Inquiries:

While Bohr's reactions to Einstein's difficulties significantly affect science and reasoning, they have not completely settled every one of the secrets of quantum mechanics. Various open inquiries and progressing discusses keep on investigating the complexities of the quantum world and its translation.

1. **Quantum Establishments:**

   Bohr's reactions brought up issues about the underpinnings of quantum mechanics, including the idea of estimation, the presence of stowed away factors, and the job of the spectator. These points remain subjects of dynamic examination and philosophical request, with suggestions for the eventual fate of quantum hypothesis.

2. **Quantum Gravity and Unification:**

The test of accommodating quantum mechanics with Einstein's hypothesis of general relativity, which depicts the gravitational power, stays a major inquiry in hypothetical physical science. Bohr's experiences into the impediments of traditional ideas might have suggestions for the improvement of a quantum hypothesis of gravity.

8.1 Discussion of the lasting impact of the Einstein-Bohr debates on physics.

The discussions between Albert Einstein and Niels Bohr, which crossed quite a few years and rotated around the translation and ramifications of quantum mechanics, lastingly affect the field of material science. These discussions, described by profound scholarly conflicts, tested tried and true way of thinking, and reshaped how we might interpret the quantum world. In this investigation, we will examine the persevering through impact of the Einstein-Bohr banters on physical science and its resulting advancement.

1. **Presentation: The Conflict of Titans**

   The Einstein-Bohr discusses, which started in the mid twentieth 100 years and went on for a long time, addressed a conflict of two titanic brains in the domain of physical science. At the core of their conflicts were major inquiries regarding the idea of the real world, determinism, and the translation of quantum mechanics. Their discussions were not simple scholarly fighting but rather significantly compelling in molding the course of physical science.

2. **Effect on Quantum Mechanics:**

The effect of the Einstein-Bohr banters on quantum mechanics, the part of physical science that arrangements with the way of behaving of particles on the littlest scales, is
critical and diverse. The discussions brought up basic issues as well as motivated further exploration and hypothetical turns of events.

1. **Characterizing the Quantum World:**

   Einstein's difficulties to the probabilistic and indeterministic nature of quantum mechanics constrained physicists to wrestle with the crucial attributes of the quantum world. Bohr's reactions, including his protection of complementarity and non-territory, characterized the characteristics of quantum reality.

2. **Molding Understandings:**

The discussions added to the improvement of different understandings of quantum mechanics. Bohr's Copenhagen understanding, which underlines the job of the spectator and the probabilistic idea of quantum results, stays quite possibly of the most generally educated translation. Einstein's investigates enlivened elective translations, for example, the pilot-wave hypothesis and the many-universes understanding, which investigate various approaches to accommodating quantum peculiarities with traditional instincts.

III. Effect on Way of thinking of Science:

The Einstein-Bohr discusses significantly affected the way of thinking of science, bringing up issues about the idea of logical speculations, the connection among hypothesis and reality, and the job of determinism in actual speculations.

1. **Authenticity versus Hostile to Authenticity:**

   The discussions prodded conversations about logical authenticity, the view that logical hypotheses give an exact portrayal of an objective reality. Einstein's suspicion about the culmination of quantum mechanics tested the pragmatist viewpoint, while Bohr's accentuation on the constraints of traditional ideas empowered enemy of authenticity, the thought that speculations may not give an exacting portrayal of the real world.

2. **Onlooker Impact and Epistemic Shift:**

Bohr's contention that the demonstration of estimation in quantum mechanics impacts the properties being noticed presented the idea of the spectator impact. This brought up issues about the epistemic idea of logical information, recommending that our perceptions are not inactive impressions of an outer reality but rather dynamic commitment with the actual world.

IV. Effect on Exploratory Material science:

The Einstein-Bohr discusses likewise substantially affected trial physical science, moving specialists to configuration explores that could test the forecasts and claims of quantum mechanics.

1. **Chime's Imbalances and Non-Area:**

   John Chime's hypothesis, formed as a reaction to the Einstein-Podolsky-Rosen (EPR) mystery, was propelled by the discussions. Chime's work demonstrated the way that no hypothesis in view of nearby secret factors could replicate the connections anticipated by quantum mechanics. Exploratory trial of Chime's imbalances affirmed the non-neighborhood nature of quantum ensnarement, lining up with Bohr's acknowledgment of non-area.

2. **Innovative Progressions:**

The discussions assumed a part in the improvement of quantum innovations. Quantum registering, quantum cryptography, and quantum instant transportation, all in light of the standards of quantum mechanics, have profited from trial checks of quantum peculiarities that were provoked, to some degree, by the discussions.

V. Effect on Ensuing Ages:

The Einstein-Bohr discusses impacted the reasoning of ensuing ages of physicists and rationalists. The discussions, alongside the positions taken by Einstein and Bohr, keep on being subjects of study, motivation, and discussion.

1. **Quantum Data Science:**

The acknowledgment of non-area, as shielded by Bohr, is basic to the field of quantum data science. Specialists in this field are investigating ways of saddling the exceptional properties of quantum mechanics, like superposition and snare, for applications in quantum figuring, quantum cryptography, and quantum correspondence.

2. **Philosophical Request:**

The discussions keep on energizing philosophical investigations into the idea of the real world and the connection among old style and quantum physical science. Scholars have investigated the ramifications of quantum mechanics for mysticism, epistemology, and the way of thinking of science, drawing on the contentions and positions set forth by Einstein and Bohr.

VI. Continuous Difficulties and Open Inquiries:

While the Einstein-Bohr discusses have had a significant and enduring effect, they have not settled every one of the secrets of the quantum world. Various difficulties and open inquiries persevere, and physicists and savants keep on investigating the complexities of quantum mechanics.

1. **Quantum Gravity and Unification:**
   The test of accommodating quantum mechanics with Einstein's hypothesis of general relativity, which depicts the gravitational power, stays a strange issue in hypothetical material science. The discussions have prodded continuous endeavors to foster a quantum hypothesis of gravity that can bind together these two major speculations.
2. **Quantum Establishments:**

The discussions brought up issues about the underpinnings of quantum mechanics, including the idea of estimation, the presence of stowed away factors, and the ramifications of non-area. These points keep on being subjects of dynamic exploration and philosophical request, with suggestions for the eventual fate of quantum hypothesis.

8.2 How their clash influenced the development of quantum physics and philosophy.

The conflict between quantum physical science and reasoning is a captivating and unpredictable episode throughout the entire existence of science and thought. It addresses a gathering point of two apparently unique domains: the exact, numerical space of quantum physical science and the theoretical, speculative universe of reasoning. This conflict has altogether affected the improvement of the two fields, molding the manner in which we grasp the principal idea of the real world. In this article, we will dive into the authentic and applied parts of this conflict, investigating how it has prompted significant changes in how we might interpret the universe and our place inside it.

1. **The Introduction of Quantum Physical science**

   To grasp the conflict between quantum physical science and reasoning, we should initially figure out the beginnings of quantum physical science. The late nineteenth and mid twentieth hundreds of years denoted an extraordinary period throughout the entire existence of material science. Old style material science, which had effectively made sense of the movement of divine bodies and the way of behaving of issue for a really long time, was confronting unconquerable difficulties in the minuscule world.

   One of the urgent minutes in this progress was the revelation of Max Planck's quantum hypothesis in 1900. Planck suggested that energy is quantized, meaning it can exist in discrete bundles, or "quanta." This progressive thought established the groundwork for quantum physical science, a field that would before long become inseparable from unusualness, vulnerability, and a take-off from traditional determinism.

2. **The Conflict Starts: Einstein versus Bohr**

   The conflict between quantum physical science and reasoning picked up speed with the discussion between two of the best personalities of the twentieth 100 years: Albert Einstein and Niels Bohr. Their scholarly showdown, which crossed a very long while, featured the philosophical ramifications of quantum mechanics.

   Einstein, well known for his hypothesis of relativity, was profoundly awkward with the probabilistic idea of quantum mechanics. He broadly pronounced, "God doesn't play dice with the universe." Einstein's position was established in his confidence in a deterministic universe, where all occasions could be anticipated with adequate information on starting circumstances. He scrutinized the culmination of quantum mechanics and looked for buried factors that would reestablish determinism.

   Bohr, then again, supported the Copenhagen understanding of quantum mechanics, which placed that the demonstration of estimation fell the quantum state into a positive result, bringing an inborn indeterminacy into the texture of the real world. Bohr contended that quantum peculiarities ought to be acknowledged for what they were, without the requirement for buried factors. He accepted that quantum mechanics gave a total depiction of the quantum world.

This conflict among Einstein and Bohr, frequently worked out in lively discussions and composed trades, brought the philosophical underpinnings of quantum mechanics to the very front of logical talk. It constrained physicists to wrestle with significant inquiries regarding the idea of the real world, the job of spectators, and the constraints of human getting it.

3. **Vulnerability Guideline and Complementarity**

At the core of the conflict between quantum material science and reasoning untruths Werner Heisenberg's Vulnerability Standard, figured out in 1927. This rule expresses that specific sets of properties, like position and energy, can't be at the same time known with erratic accuracy. The more precisely one property is known, the less precisely the other not entirely set in stone.

The Vulnerability Rule had significant philosophical ramifications. It tested the old style thought of a deterministic, perfect timing universe, where everything could be unequivocally estimated and anticipated. All things considered, it acquainted an intrinsic

cutoff with our insight, recommending that there were principal cutoff points to what we could be aware of the quantum world.

Bohr further fostered these thoughts through his idea of "complementarity." As per Bohr, different trial arrangements could uncover various parts of a quantum framework's way of behaving. For instance, an examination intended to gauge an electron's molecule like properties wouldn't at the same time uncover its wave-like properties. This thought of complementarity underlined that the depiction of quantum peculiarities relied upon the exploratory setting, obscuring the line between true reality and the spectator's point of view.

4. **Quantum Mechanics and Philosophical Understandings**
The conflict between quantum material science and reasoning led to different philosophical understandings of quantum mechanics. These translations looked to get a handle on the hypothesis' suggestions for the idea of the real world and the job of cognizance in the estimation cycle.

Copenhagen Understanding: Niels Bohr's Copenhagen translation stayed one of the prevailing perspectives. It accentuated the job of the spectator and the wave capability breakdown during estimation. As indicated by this understanding, quantum mechanics gave a total portrayal of the quantum world, yet the demonstration of estimation presented indeterminacy.

Many-Universes Translation: Proposed by Hugh Everett III during the 1950s, the Many-Universes understanding recommends that the universe parts into different branches with every quantum occasion, making a tremendous multiverse of equal real factors. In this view, all potential results of quantum occasions happen in discrete branches, staying away from the requirement for wave capability breakdown.

Pilot-Wave Hypothesis: Louis de Broglie and David Bohm fostered the pilot-wave hypothesis, which places that particles are directed by stowed away pilot waves that decide their way of behaving. This understanding jelly determinism however presents non-nearby secret factors.

Objective Breakdown Models: A few understandings recommend that the wave capability falls precipitously, without the requirement for an onlooker. These models expect to determine the estimation issue by presenting new physical science that triggers wave capability breakdown.

These understandings delineate how quantum material science started extreme philosophical discussions, with every translation offering a novel point of view on the idea of the real world and the eyewitness' part in quantum peculiarities.

5. **Suggestions for Theory**

The conflict between quantum physical science and reasoning had extensive ramifications for philosophical idea past the limits of the quantum domain. It tested a portion of the essential presumptions that had supported Western way of thinking for a really long time.

Epistemological Shift: Quantum mechanics presented a change in epistemology, the part of reasoning worried about information. The Vulnerability Standard proposed that there were innate cutoff points to what we could be familiar with the actual world, testing the old style thought of outright assurance in information.

Ontological Ramifications: The philosophical understandings of quantum mechanics brought up issues about the idea of reality itself. The presence of equal universes in the Many-Universes understanding and the job of stowed away factors in pilot-wave hypothesis constrained rationalists to wrestle with the idea of a complex reality.

Spectator Impact: The job of the onlooker in quantum peculiarities brought up issues about the connection among cognizance and the actual world. A few savants investigated the possibility that cognizance assumed a principal part in the development of the real world, testing the realist perspective.

Complementarity and Subjectivity: Bohr's idea of complementarity featured the abstract idea of perception. This thought tested the idea of an objective reality that exists freely of the eyewitness' viewpoint.

6. **Quantum Physical science and Contemporary Way of thinking**

The conflict between quantum physical science and reasoning keeps on affecting contemporary way of thinking. It has added to the improvement of new philosophical points of view and areas of request.

Quantum Mysticism: A few contemporary savants have dug into quantum transcendentalism, investigating the ontological ramifications of quantum mechanics. They look at inquiries concerning the idea of the real world, the job of eyewitnesses, and the connection between the quantum and old style universes.

Quantum Morals: The conflict has likewise motivated conversations about quantum morals, zeroing in on the moral ramifications of a universe that is intrinsically unsure

and uncertain. This part of reasoning investigates the moral components of vulnerability, possibility, and the restrictions of human information.

Reasoning of Science: Quantum physical science has reshaped the way of thinking of science, prompting conversations about logical authenticity, the idea of logical speculations, and the job of reasoning in deciphering logical ideas.

8.2 How their clash influenced the development of quantum physics and philosophy.

The improvement of quantum physical science and reasoning stands as a demonstration of the unique exchange between logical disclosure and philosophical thought. The conflict between these two scholarly areas has driven the progression of quantum physical science as well as reshaped the philosophical scene. In this exposition, we will investigate the significant impact of this conflict on the advancement of both quantum physical science and reasoning, featuring the manners by which it has tested existing ideal models and developed how we might interpret the central idea of the real world.

1. The Development of Quantum Material science

To see the value in how the conflict between quantum material science and reasoning has formed the two fields, we should initially dive into the beginnings of quantum physical science. In the late nineteenth and mid twentieth hundreds of years, the predominant logical perspective was established in traditional physical science, which had effectively made sense of the way of behaving of plainly visible articles for quite a long time. Be that as it may, old style material science confronted inconceivable difficulties while endeavoring to portray the way of behaving of particles at the nuclear and subatomic levels.

1. The Quantum Upset

The introduction of quantum physical science can be followed back to Max Planck's momentous work in 1900, when he presented the idea of quantization of energy. He recommended that energy isn't ceaseless yet exists in discrete bundles or "quanta." This idea denoted an extreme takeoff from traditional physical science, where energy was believed to be nonstop and vastly separable.

Planck's quantization of energy established the groundwork for quantum mechanics, another hypothetical system that looked to make sense of the way of behaving of particles at the quantum level. Key figures like Albert Einstein, Niels Bohr, Werner Heisenberg, Erwin Schrödinger, and Max Conceived would proceed to create and refine quantum mechanics, introducing a logical insurgency of remarkable extents.

II. The Philosophical Test

The rise of quantum mechanics delivered significant philosophical difficulties. Quantum peculiarities were in a general sense not the same as traditional material science, presenting ideas like wave-molecule duality, superposition, and trap, which resisted old style instincts and brought up philosophical issues about the idea of the real world, determinism, and the job of the onlooker.

1. **Einstein's Study**

   One of the most conspicuous conflicts between quantum material science and reasoning arose as Albert Einstein's suspicion. Einstein was profoundly grieved by the probabilistic idea of quantum mechanics, broadly expressing, "God doesn't play dice with the universe." He contended that the hypothesis was fragmented and that secret factors could represent the clear indeterminacy of quantum occasions.

   Einstein's study was an impetus for philosophical conversations inside established researchers. It provoked physicists and logicians the same to wrestle with inquiries regarding the culmination of quantum mechanics and the idea of determinism in the quantum world.

2. **Bohr's Reaction**

Niels Bohr, a vital figure in the improvement of quantum mechanics, offered a contrast to Einstein's complaints. He advocated the Copenhagen understanding of quantum mechanics, which underlined the job of the eyewitness and the idea of wave capability breakdown during estimation. As per Bohr, quantum mechanics gave a total portrayal of the quantum world, and the demonstration of estimation presented indeterminacy.

Bohr's reaction set up for an extended and logically rich discussion between the two monsters of science. This conflict featured the philosophical underpinnings of quantum mechanics and pushed the limits of how we might interpret the universe.

III. Heisenberg's Vulnerability Rule and Complementarity

Werner Heisenberg's Vulnerability Rule, figured out in 1927, further escalated the philosophical ramifications of quantum mechanics. This rule expresses that specific sets of properties, like position and energy, can't be all the while known with erratic accuracy. The more precisely one property is known, the less precisely the other not entirely set in stone.

1. **Cutoff points of Information**
   The Vulnerability Standard acquainted a principal limit with human information about the quantum world. It tested the traditional idea of a deterministic, perfect timing universe, where all occasions could be exactly estimated and anticipated. All things considered, it proposed that there were intrinsic cutoff points to what we could be aware of the quantum domain, developing the philosophical investigation into the idea of information itself.

2. **Bohr's Complementarity**

Niels Bohr's idea of complementarity further added to the philosophical talk encompassing quantum mechanics. Bohr contended that different exploratory arrangements could uncover various parts of a quantum framework's way of behaving. For example, an investigation intended to quantify an electron's molecule like properties wouldn't at the same time uncover its wave-like properties. This thought of complementarity underlined that the depiction of quantum peculiarities relied upon the exploratory setting, obscuring the line between genuine reality and the eyewitness' viewpoint.

IV. Philosophical Translations of Quantum Mechanics

The conflict between quantum physical science and reasoning prompted the improvement of different philosophical understandings of quantum mechanics, each offering a particular point of view on the idea of the real world and the job of perception.

1. **Copenhagen Translation**
   The Copenhagen translation, created by Bohr and his partners, stayed one of the predominant perspectives. It stressed the job of the spectator and the wave capability breakdown during estimation. As per this understanding, quantum mechanics gave a total portrayal of the quantum world, yet the demonstration of estimation presented indeterminacy.

2. **Many-Universes Understanding**

The Many-Universes understanding, proposed by Hugh Everett III during the 1950s, recommends that the universe parts into various branches with every quantum occasion, making a tremendous multiverse of equal real factors. In this view, all potential results of quantum occasions happen in discrete branches, keeping away from the requirement for wave capability breakdown. This translation tested regular ideas of reality by proposing the presence of incalculable equal universes.

3. **Pilot-Wave Hypothesis**

Louis de Broglie and David Bohm fostered the pilot-wave hypothesis, which sets that particles are directed by stowed away pilot waves that decide their way of behaving.

This translation jam determinism however presents non-neighborhood stowed away factors, testing the probabilistic idea of standard quantum mechanics.

4. **Objective Breakdown Models**

A few understandings recommend that the wave capability implodes immediately, without the requirement for an onlooker. These models expect to determine the estimation issue by presenting new material science that triggers wave capability breakdown.

These understandings highlight the extravagance of the philosophical talk started by quantum mechanics. They welcome consideration on the idea of the real world, the connection among spectators and the noticed, and the constraints of human getting it.

V. Philosophical Ramifications

The conflict between quantum physical science and reasoning has had broad ramifications for philosophical idea past the domain of quantum mechanics.

1. **Epistemological Shift**

Quantum mechanics provoked a change in epistemology — the

part of reasoning worried about information. The Vulnerability Standard and the constraints of information tested the old style idea of outright conviction in information, stressing the inborn vulnerability and limits of human comprehension.

2. **Ontological Investigation**

Logicians and physicists the same have taken part in ontological investigation, contemplating the idea of reality itself. Quantum mechanics has tested traditional ideas of a solitary, objective reality by recommending the presence of equal universes and the job of stowed away factors.

3. **Spectator Impact**

The job of the spectator in quantum peculiarities brought up issues about the connection among cognizance and the actual world. A few scholars have investigated the possibility that cognizance assumes a key part in the development of the real world, testing realist perspectives.

4. **Complementarity and Subjectivity**

Bohr's idea of complementarity featured the emotional idea of perception, obscuring the line between genuine reality and the eyewitness' viewpoint. This idea has significant ramifications for the way of thinking of discernment and the idea of subjectivity.

VI. Quantum Material science and Contemporary Way of thinking

The conflict between quantum material science and reasoning keeps on impacting contemporary philosophical idea, leading to new viewpoints and areas of request.

1. **Quantum Mysticism**

Contemporary scholars have dug into quantum power, investigating the ontological ramifications of quantum mechanics. They inspect inquiries regarding the idea of the real world, the job of onlookers, and the connection between the quantum and old style universes.

2. **Quantum Morals**

The conflict has likewise enlivened conversations about quantum morals, zeroing in on the moral ramifications of a universe that is innately dubious and uncertain. This part of reasoning investigates the moral elements of vulnerability, possibility, and the constraints of human information.

3. **Theory of Science**

Quantum physical science has reshaped the way of thinking of science, prompting conversations about logical authenticity, the idea of logical hypotheses, and the job of reasoning in deciphering logical ideas.

8.3 The broader implications of their rivalry for science and philosophy.

The contention among science and reasoning, however not a new peculiarity, has had sweeping ramifications for the two disciplines. This powerful relationship, set apart by participation, rivalry, and strain, has altogether formed the advancement of science and the development of philosophical idea. In this article, we will investigate the more extensive ramifications of their competition, revealing insight into what it has meant for not just the separate areas of science and reasoning yet additionally how we might interpret the world, human information, and the idea of the real world.

1. **The Exchange of Science and Reasoning**

To comprehend the ramifications of the contention among science and reasoning, perceiving their interchange from the beginning of time is essential. In old times, reasoning enveloped many requests, including those we presently consider logical. In any case, as observational strategies and particular areas of science arose, a disparity happened.

1. **The Ascent of Experimental Science**
   During the Logical Insurgency of the seventeenth hundred years, figures like Galileo Galilei and Isaac Newton presented thorough exact techniques, numerical accuracy, and trial confirmation into the investigation of the regular world. This shift prompted the foundation of science as an unmistakable discipline, zeroing in on the deliberate investigation of actual peculiarities.

2. **Theory's Advancing Job**

In the interim, reasoning adjusted to its evolving job. It moved from including all parts of information to looking at key inquiries concerning presence, morals, epistemology, and mysticism. As science thrived, reasoning wound up managing inquiries past exact examination, like morals, meaning, and the idea of the real world.

II. The Serious Strain: Reductionism versus Comprehensive quality

One of the persevering through subjects in the contention among science and reasoning is the strain among reductionism and comprehensive quality. Reductionism looks to figure out complex peculiarities by separating them into less complex parts, while comprehensive quality underscores the interconnectedness of frameworks and the significance of thinking about the entirety.

1. **Science's Hug of Reductionism**
   Current science has frequently embraced reductionism as a strong way to deal with grasping the normal world. Fields like physical science, science, and atomic science have made exceptional progress by diminishing complex peculiarities to their constituent parts. This approach has yielded significant bits of knowledge and mechanical progressions.

2. **Theory's All encompassing Point of view**

Conversely, theory has habitually taken a more all encompassing viewpoint. Scholars have investigated inquiries regarding the idea of awareness, morals, and the human experience by considering the more extensive setting where these peculiarities happen. Comprehensive quality in way of thinking empowers the assessment of mind boggling interrelationships and developing properties.

III. Suggestions for Information and Understanding

The competition among science and reasoning has significant ramifications for human information and our mission to grasp the world.

1. **Cutoff points of Logical Reductionism**

   While reductionism has been instrumental in progressing logical information, it has constraints. Reductionist methodologies might battle to catch the extravagance and intricacy of specific peculiarities, especially in fields like cognizance studies and morals. This restriction highlights the significance of reasoning in resolving inquiries past the scope of reductionist science.

2. **The Correlative Idea of Disciplines**

The contention has featured the corresponding idea of science and reasoning. They act as two focal points through which we can inspect the world. Science gives experimental proof and testable theories, while theory offers theoretical systems and basic investigation. Together, they add to a more far reaching comprehension of the real world.

IV. Moral and Moral Contemplations

The competition among science and reasoning has moral and moral ramifications. The two disciplines assume a part in forming our qualities and directing moral navigation.

1. **Moral Appearance in Way of thinking**

   Reasoning has for quite some time been worried about morals, investigating inquiries regarding ethical quality, equity, and easy street. Logicians take part in moral reflection, giving significant direction to people and social orders in exploring complex moral difficulties.

2. **Moral Aspects in Science**

Science additionally wrestles with moral contemplations, particularly in regions like bioethics and natural morals. Moral worries emerge when logical progressions can possibly affect human lives, the climate, or people in the future. Reasoning's moral experiences assist with illuminating mindful logical direct.

V. The Idea of The real world and Presence

The contention among science and reasoning has significant ramifications for how we might interpret the idea of the real world and presence.

1. **Science's Investigation of the Actual Universe**

   Science has taken critical steps in disclosing the actual universe's activities, from the major particles of molecule physical science to the enormous spans investigated by astronomy. These logical disclosures have reclassified how we might interpret the actual universe and its starting points.

2. **Theory's Assessment of Presence**

Reasoning, then again, dives into inquiries of presence and reality that rise above the observational. Logicians ponder the idea of cognizance, the presence of unique elements, and the significance of life. In doing as such, they add to how we might interpret presence from a more extensive perspective.

VI. The Job of Wariness and Decisive Reasoning

The contention among science and reasoning highlights the significance of doubt and decisive reasoning in the two disciplines.

1. **Logical Doubt**

   Science depends on doubt as a key guideline. Researchers

question suppositions, challenge speculations, and thoroughly test claims. This distrust defends the trustworthiness of logical request and advances the quest for exact truth.

2. Philosophical Decisive Reasoning

Theory likewise puts a superior on decisive reasoning. Thinkers break down contentions, survey presumptions, and investigate elective perspectives. This decisive reasoning is fundamental for developing sound moral structures, assessing complex magical ideas, and resolving existential inquiries.

VII. Overcoming any issues: Interdisciplinary Methodologies

Perceiving the limits of a selective contention among science and reasoning, interdisciplinary methodologies have gotten forward movement.

1. Reasoning of Science

The way of thinking of science fills in as a scaffold between the two disciplines. It analyzes the idea of logical request, the design of logical speculations, and the connection among science and different subject matters. Thinkers of science help to explain the establishments and limits of logical practice.

2. Moral Ramifications in Science

Interdisciplinary coordinated effort among ethicists and researchers has become progressively significant, especially in fields like biotechnology and man-made brainpower. Ethicists add to the improvement of capable logical practices and assist with tending to moral issues raised by mechanical progressions.

3. Mental Science and Awareness Studies

Mental science, which joins experiences from brain research, neuroscience, and reasoning, investigates questions connected with cognizance, discernment, and the psyche. This interdisciplinary methodology looks to overcome any issues between logical examination and philosophical request in grasping the idea of awareness.

VIII. The Contention's Significance in Contemporary Discussions

The contention among science and reasoning remaining parts pertinent in contemporary discussions on issues, for example, man-made consciousness, environmental change, morals in biotechnology, and the underpinnings of quantum mechanics.

1. **Moral Contemplations in computer based intelligence**

   The improvement of man-made brainpower (artificial intelligence) brings up significant moral issues about independence, obligation, and the likely results of artificial intelligence frameworks. Rationalists and ethicists assume a crucial part in directing conversations on the moral components of artificial intelligence.

2. **Environmental Change and Natural Morals**

   Environmental change discusses envelop logical evaluations, strategy choices, and moral contemplations about our obligation to people in the future and the climate. Natural morals, a part of reasoning, illuminates conversations on maintainability and biological stewardship.

3. **Quantum Mechanics and Reality**

The conflict between quantum material science and reasoning keeps on rousing conversations about the idea of the real world and the job of perception. Scholars wrestle with the ramifications of quantum mechanics for how we might interpret the universe.

Chapter 9

Beyond Einstein and Bohr

The names Albert Einstein and Niels Bohr are inseparable from the logical upheavals of the mid twentieth hundred years. Their noteworthy commitments to material science, especially in the domains of relativity and quantum mechanics, have molded how we might interpret the universe. Notwithstanding, the tradition of these two monsters reaches out past their singular accomplishments. This exposition investigates the improvements in present day physical science and quantum mechanics that have happened right after Einstein and Bohr, revealing insight into how their work established the groundwork for additional headways and made the way for new boondocks in science.

1. **Einstein's Relativity and Its Continuous Effect**
1. **Unique Relativity**
   Albert Einstein's 1905 hypothesis of unique relativity changed how we might interpret reality. Exceptional relativity presented the idea that the laws of physical science are no different for all eyewitnesses, no matter what their relative movement. This standard prompted the renowned condition $E=mc^2$, which uncovers the equality of mass and energy.

Influence on Material science: Extraordinary relativity tested old style thoughts of outright existence, generally changing comprehension we might interpret the universe. It laid the foundation for the improvement of molecule material science, cosmology, and the investigation of high-energy peculiarities.

Mechanical Applications: Unique relativity has commonsense applications in current innovation, especially in the plan of molecule gas pedals and GPS frameworks, where relativistic impacts should be considered.

2. General Relativity

Einstein's overall hypothesis of relativity, distributed in 1915, went further by depicting gravity as the twisting of spacetime by mass and energy. This progressive hypothesis anticipated the twisting of light by gravity, known as gravitational lensing, and the presence of dark openings.

Influence on Astronomy: General relativity gave the hypothetical system to grasping the way of behaving of gigantic articles in the universe. It has been instrumental in the investigation of the extending universe, the arrangement of worlds, and the discovery of gravitational waves.

Testing and Affirmation: General relativity has gone through thorough exploratory testing, with perceptions of peculiarities, for example, the precession of Mercury's circle and the twisting of starlight during sun based shrouds affirming its expectations.

II. Quantum Mechanics and the Tradition of Bohr

Niels Bohr's work in the early improvement of quantum mechanics, especially his Copenhagen translation, denoted a huge takeoff from old style material science. While Bohr's thoughts lighted discussions with Einstein, they established the groundwork for a quantum insurgency that keeps on developing.

1. **Quantum Mechanics**

   Quantum mechanics, as formalized by Bohr, Werner Heisenberg, Erwin Schrödinger, and Max Conceived, presented the idea of quantized energy levels, wave-molecule duality, and the probabilistic idea of particles at the quantum level.

   The Copenhagen Translation: Bohr's Copenhagen understanding accentuated the job of the eyewitness and the idea of wave capability breakdown during estimation. It prompted banters about the idea of the real world and the spectator's effect on quantum peculiarities.

   Quantum Trap: Einstein, alongside Boris Podolsky and Nathan Rosen, tested quantum mechanics through the EPR oddity, which featured the peculiarity of quantum snare, where the properties of two particles become associated in any event, when isolated by immense distances.

2. **The Quantum Upheaval**

The tradition of Bohr's work stretches out into the continuous quantum transformation, set apart by progressions in quantum processing, quantum cryptography, and quantum data hypothesis.

Quantum Processing: Quantum PCs tackle the standards of quantum mechanics to play out specific computations dramatically quicker than traditional PCs. Organizations

like IBM, Google, and Microsoft are effectively seeking after quantum processing research.

Quantum Cryptography: Quantum cryptography uses the standards of quantum ensnarement to make secure correspondence channels. This innovation holds the commitment of strong encryption and has suggestions for network safety.

Quantum Data Hypothesis: Quantum data hypothesis investigates the central idea of data in quantum frameworks, prompting the improvement of quantum calculations and the investigation of quantum entrapment as an asset for correspondence and calculation.

III. Past Einstein: New Skylines in Physical science

1. Brought together Speculations

Einstein spent a lot of his later vocation in quest for a brought together hypothesis that would accommodate general relativity with quantum mechanics. While he didn't prevail in the course of his life, his journey has motivated continuous endeavors to bring together the central powers of the universe.

String Hypothesis: String hypothesis is a possibility for a bound together hypothesis of material science. It places that the crucial structure blocks of the universe are not particles but rather minuscule vibrating strings. String hypothesis has various varieties, including superstring hypothesis and M-hypothesis, which intend to bind together the powers of nature.

Fabulous Bound together Speculations (GUTs): GUTs look to bind together the electromagnetic, feeble, and solid atomic powers into a solitary system. While these hypotheses have gained huge headway in making sense of the way of behaving of particles, exploratory confirmation stays a test.

2. Dull Matter and Dim Energy

The secrets of dull matter and dim energy address the absolute most critical riddles in current cosmology. These mysterious substances, which don't connect with light or customary matter, have been the focal point of broad exploration.

Dull Matter: Dim matter is accepted to make up a significant piece of the universe's mass, yet it stays imperceptible through customary means. Researchers are directing tests and perceptions to more readily grasp its temperament.

Dull Energy: Dim energy is liable for the sped up extension of the universe, a revelation made in the late twentieth 100 years. Its real essence stays slippery, yet it challenges how we might interpret the universe's definitive destiny.

3. Multiverse Hypotheses

Multiverse hypotheses propose the presence of different universes past our detectable universe. While they are speculative, these hypotheses stand out in the journey to make sense of the tweaking of actual constants.

Many-Universes Understanding: In quantum mechanics, the Many-Universes translation proposes that each quantum occasion brings about an expanding of the universe, prompting a large number of equal real factors. This understanding brings up issues about the idea of the real world and the job of perception.

Vast Expansion and Air pocket Universes: A few cosmological models, like infinite expansion hypothesis, propose the quick extension of the universe led to "bubble universes." Each air pocket could have different actual constants, prompting a multiverse.

IV. Quantum Mechanics in the 21st 100 years

Quantum mechanics keeps on being a lively and developing field of study with various ramifications for innovation, calculation, and our comprehension of the universe.

1. **Quantum Processing Headways**

   Progressions in quantum figuring can possibly alter fields like cryptography, materials science, drug disclosure, and improvement issues. Specialists are attempting to construct more steady and versatile quantum PCs.

2. **Quantum Correspondence and Cryptography**

   Quantum correspondence innovations, including quantum key dissemination, offer secure method for sending data. Quantum cryptography research investigates novel strategies for encryption and secure correspondence.

3. **Quantum Ensnarement and Ringer Tests**

   Trial of quantum ensnarement, for example, Chime tests, keep on testing how we might interpret quantum physical science and the idea of the real world. These investigations investigate the infringement of Ringer imbalances and the constraints of neigh-borhood authenticity.

4. **Quantum Data and Quantum Gravity**

Quantum data hypothesis meets with the investigation of quantum gravity, meaning to accommodate quantum mechanics with the gravitational power. This try has prompted examinations concerning the idea of spacetime at the quantum level.

V. Theory's Proceeded with Importance

While the competition among science and reasoning might have developed, reasoning remaining parts an imperative part of our journey to figure out the ramifications of logical headways and the central inquiries of presence.

1. **Reasoning of Science**
   The way of thinking of science keeps on analyzing the idea of logical speculations, the division issue (recognizing science from pseudoscience), and the moral ramifications of logical exploration.

2. **Morals and Innovation**
   Thinkers assume a urgent part in resolving moral inquiries raised by new advancements, from hereditary designing and man-made reasoning to biotechnology and natural preservation.

3. **Mysticism and Reality**

Scholars dig into inquiries concerning the idea of the real world, the psyche body issue, and the presence of unique elements. These requests meet with advancements in physical science, like the ramifications of quantum mechanics for how we might interpret awareness.

9.1A look at the subsequent developments in quantum physics.

Quantum material science, conceived out of the turbulent mid twentieth hundred years with figures like Max Planck, Albert Einstein, and Niels Bohr, has proceeded to develop and shape how we might interpret the central idea of the universe. Past its central standards, resulting improvements in quantum physical science have extended the field into energizing new wildernesses, testing our instincts and promising progressive advancements. In this exposition, we will investigate a portion of the huge turns of events and forward leaps that have happened in quantum physical science since its commencement.

1. **Quantum Mechanics and the Standard Model**
1. **The Introduction of Quantum Mechanics**

   Quantum mechanics, as formalized by Bohr, Heisenberg, Schrödinger, and Conceived, presented the probabilistic idea of particles at the quantum level, wave-molecule duality, and the idea of quantized energy levels. This changed comprehension we might interpret the way of behaving of particles and established the groundwork for resulting advancements.

   Wave Capabilities and Quantum Expresses: Schrödinger's wave condition gave a numerical structure to portraying quantum states, addressed by wave capabilities. These wave capabilities encode data about a molecule's situation, force, and energy.

   The Vulnerability Guideline: Heisenberg's Vulnerability Rule, formed in 1927, laid out a principal cutoff to the accuracy with which certain sets of properties, like position and energy, could be all the while known.

2. **The Standard Model of Molecule Physical science**

The Standard Model is the predominant hypothetical system in molecule material science, binding together three of the four major powers of nature (electromagnetism, the feeble atomic power, and the solid atomic power) and depicting the way of behaving of rudimentary particles.

Electroweak Hypothesis: Sheldon Glashow, Abdus Salam, and Steven Weinberg fostered the electroweak hypothesis, which binds together electromagnetism and the feeble atomic power. This hypothesis was affirmed through tests, prompting the honor of the Nobel Prize in Material science in 1979.

Disclosure of the Higgs Boson: In 2012, the Huge Hadron Collider (LHC) at CERN affirmed the presence of the Higgs boson, a molecule related with the Higgs field, which is liable for giving mass to different particles. This disclosure finished the Standard Model.

II. Quantum Entrapment and Ringer's Hypothesis

Quantum entrapment, a peculiarity where the properties of at least two particles become corresponded, in any event, when isolated by enormous distances, has kept on being a subject of extraordinary review and trial and error.

1. **Einstein-Podolsky-Rosen (EPR) Catch 22**

   In 1935, Albert Einstein, Boris Podolsky, and Nathan Rosen distributed a paper presenting the EPR Catch 22. They contended that quantum mechanics may be deficient in light of the fact that it apparently considered quick "creepy activity a ways off." This mystery tested the idea of neighborhood authenticity, which places that far off occasions can't impact each other immediately.

2. **Chime's Hypothesis**

   In 1964, physicist John Chime proposed a hypothesis that could tentatively test the forecasts of quantum mechanics against neighborhood authenticity. Chime's examinations, led during the 1980s and 1990s, reliably affirmed quantum forecasts and precluded neighborhood stowed away factor hypotheses.

3. **Quantum Data and Quantum Cryptography**

Quantum ensnarement has down to earth applications in the field of quantum data and quantum cryptography. Analysts have outfit ensnared particles for secure correspondence and quantum figuring.

Quantum Key Conveyance (QKD): Quantum cryptography conventions like QKD use the standards of ensnarement to make rugged encryption keys. These frameworks are viewed as profoundly secure and can possibly alter information security.

Quantum Instant transportation: Quantum instant transportation is an interaction that permits the quantum condition of one molecule to be sent to another, far off molecule. It has applications in quantum correspondence and quantum processing.

III. Quantum Processing and Quantum Calculations

Quantum processing has arisen as quite possibly of the most encouraging mechanical improvement lately. Quantum PCs influence the remarkable properties of quantum mechanics to play out specific estimations dramatically quicker than old style PCs.

1. **Quantum Pieces (Qubits)**

   Old style PCs use bits as the major units of data, addressed as 0 or 1. Quantum PCs use qubits, which can exist in a superposition of both 0 and 1 states at the same time, dramatically expanding their computational power.

2. **Shor's Calculation**

   Peter Shor's calculation, created in 1994, showed the way that quantum PCs could proficiently factor huge numbers, an issue considered immovable for traditional PCs. Shor's calculation has significant ramifications for cryptography, as it might actually break broadly utilized encryption techniques.

3. **Grover's Calculation**

   Lov Grover's calculation, created in 1996, demonstrated the way that quantum PCs could perform unstructured pursuit undertakings quadratically quicker than old style PCs. This calculation has applications in data set search and enhancement issues.

4. **Quantum Incomparability**

In 2019, Google professed to have accomplished quantum matchless quality, showing the way that its quantum PC, Sycamore, could play out a particular estimation quicker than the world's most remarkable traditional supercomputers. This undeniable a huge achievement in the improvement of quantum figuring.

IV. Quantum Examinations and Crucial Tests

Quantum material science has been dependent upon progressively exact analyses and tests, frequently testing our instinctive comprehension of the quantum world.

1. **Quantum Decoherence**

   Quantum decoherence alludes to the peculiarity where quantum frameworks lose their quantum properties and act traditionally while interfacing with their current circumstance. Specialists have been exploring ways of controlling and alleviate decoherence to keep up with the fragile quantum states important for quantum advances.

2. **Quantum Incomparability and Intricacy**

   Hypothetical physicists have been investigating the computational furthest reaches of quantum PCs and the intricacy of quantum calculations. Understanding these cutoff points is essential for tackling the maximum capacity of quantum processing.

3. **Quantum Recreations**

Quantum PCs can possibly mimic quantum frameworks precisely, which is trying for old style PCs. This capacity can support figuring out complex quantum peculiarities, like high-temperature superconductivity or the way of behaving of unequivocally associated electrons.

V. Quantum Advances and Applications

The headways in quantum material science have led to a variety of arising advances and viable applications.

1. **Quantum Detecting**

   Quantum sensors, like nuclear timekeepers and magnetometers, offer unrivaled accuracy in estimating different actual amounts.

They have applications in route, topography, and principal material science research.

2. **Quantum Imaging**

Quantum imaging methods, similar to quantum-upgraded cameras, empower the location of weak signals and further developed imaging in low-light circumstances. These advances have applications in clinical imaging and remote detecting.

3. **Quantum Correspondence**

Quantum correspondence frameworks, including quantum key circulation, can possibly give secure correspondence channels insusceptible to snoopping. These frameworks are now being sent for secure information transmission.

4. **Quantum Metrology**

Quantum metrology utilizes quantum standards to upgrade estimation accuracy. It is fundamental in fields like gravitational wave identification and the improvement of new principles for estimations.

VI. The Continuous Mission for Quantum Gravity

Quite possibly of the most significant test in current material science is the journey for a hypothesis of quantum gravity, which would bring together quantum mechanics and general relativity. Such a hypothesis would give an exhaustive structure to figuring out the way of behaving of issue and energy at all sizes of the universe.

1. **String Hypothesis and Then some**

String hypothesis recommends that the central structure blocks of the universe are not particles but rather little vibrating strings. It tries to bring together the powers of nature, including gravity, inside a solitary hypothetical structure.

2. **Circle Quantum Gravity**

Circle quantum gravity is one more way to deal with quantum gravity that treats reality as quantized substances. It means to accommodate general relativity with quantum mechanics by quantizing the calculation of spacetime.

3. **Dark Opening Data Oddity**

The investigation of dark openings keeps on being a fruitful ground for investigating the interchange between quantum mechanics and gravity. The goal of the dark opening data oddity, which questions the conservation of data inside dark openings, stays an open inquiry.

VII. The Eventual fate of Quantum Material science

The field of quantum physical science is ready for additional development and investigation in the next few decades. A few vital areas of interest include:

1. **Quantum Materials**
   Quantum materials, for example, high-temperature superconductors and topological separators, offer novel properties that could change hardware, energy capacity, and quantum processing.

2. **Quantum Science**
   Quantum peculiarities might assume a part in natural cycles, like photosynthesis and avian route. Specialists are exploring the way in which quantum impacts impact organic frameworks.

3. **Quantum Man-made brainpower**
   Quantum registering holds the possibility to fundamentally speed up AI calculations, prompting progressions in computerized reasoning and information examination.

4. **Quantum Organizations**

The improvement of quantum organizations, fit for sending quantum data over significant distances, could empower secure correspondence and conveyed quantum registering.

9.2 The contributions of other physicists in shaping the field.

While Albert Einstein and Niels Bohr are much of the time celebrated as the monsters of twentieth century material science, it's fundamental to recognize that they were important for a more extensive local area of splendid personalities. Various different physicists made momentous commitments that assumed a significant part in molding the field of material science. In this article, we will investigate the commitments of a portion of these uncelebrated yet truly great individuals whose work fundamentally progressed how we might interpret the universe.

1. Max Planck: Father of Quantum Hypothesis

Before the appearance of quantum mechanics, traditional physical science couldn't make sense of specific peculiarities, like the way of behaving of blackbody radiation. Max Planck, a German physicist, made a strong stride by presenting the idea of quantization of energy.

1. **Planck's Quantum Speculation**
   In 1900, Planck suggested that energy is quantized, meaning it can exist in discrete units or "quanta." He presented the Planck steady (h) to depict this quantization. This momentous thought established the groundwork for quantum mechanics.
2. **Birth of Quantum Mechanics**

Planck's work was the impetus for the advancement of quantum mechanics, as later refined by Bohr, Heisenberg, Schrödinger, and Conceived. His commitments procured him the Nobel Prize in Physical science in 1918.

II. Werner Heisenberg: Vulnerability Rule

Werner Heisenberg, a German physicist, made a significant commitment to quantum mechanics with his detailing of the Vulnerability Standard, which tested the thought of exact estimation in the quantum world.

1. **The Vulnerability Rule**
   In 1927, Heisenberg recommended that it is difficult to gauge

both the position and energy of a quantum molecule with erratic accuracy all the while. This guideline uncovered a major breaking point as far as anyone is concerned of quantum frameworks.

2. **Lattice Mechanics**

Heisenberg's lattice mechanics, created in lined up with Schrödinger's wave mechanics, gave another numerical structure to quantum material science. This network approach was instrumental in improving on estimations and figuring out quantum peculiarities.

III. Erwin Schrödinger: Wave Condition

Erwin Schrödinger, an Austrian physicist, presented the Schrödinger condition in 1926, a vital condition in quantum mechanics that depicts the way of behaving of quantum frameworks.

1. **Wave Mechanics**

   Schrödinger's wave condition presented the idea of wave capabilities, which address the likelihood dispersions of particles in quantum frameworks. Wave mechanics gave a strong and natural system for figuring out quantum conduct.

2. **Nobel Prize**

Schrödinger's work acquired him the Nobel Prize in Physical science in 1933, close by Paul Dirac, for his advancement of the wave condition.

IV. Paul Dirac: Dirac Condition and Antimatter

Paul Dirac, an English physicist, made huge commitments to the improvement of quantum mechanics, especially with his detailing of the Dirac condition and the expectation of antimatter.

1. **Dirac Condition**

   In 1928, Dirac formed the Dirac condition, which brought together quantum mechanics with extraordinary relativity. This condition portrayed the way of behaving of electrons moving at relativistic rates and anticipated the presence of particles with both positive and negative energies.

2. Revelation of Antimatter

Dirac's condition anticipated the presence of antimatter particles, like the positron (the antimatter partner of the electron). This expectation was affirmed tentatively via Carl Anderson in 1932, denoting a progressive disclosure in molecule material science.

V. Enrico Fermi: Trailblazer of Atomic Physical science

Italian physicist Enrico Fermi made historic commitments to atomic physical science, molecule physical science, and the improvement of the principal atomic reactor.

1. Fermi's Hypothesis of Beta Rot

Fermi proposed a hypothesis of beta rot in 1934, making sense of how shaky nuclear cores emanate electrons or positrons. This work established the groundwork for the investigation of feeble atomic collaborations and the comprehension of the powerless power.

2. Fermi's Atomic Reactor

During The Second Great War, Fermi and his group at the College of Chicago accomplished the principal controlled atomic chain response in 1942, denoting a pivotal achievement in the improvement of thermal power and the nuclear bomb.

VI. Subrahmanyan Chandrasekhar: Heavenly Development

Indian-American astrophysicist Subrahmanyan Chandrasekhar made critical commitments to how we might interpret heavenly development, especially the destiny of monstrous stars.

1. Chandrasekhar Cutoff

Chandrasekhar proposed the idea of as far as possible in 1930, which portrays the most extreme mass of a white small star past which it would implode into a neutron star or dark opening. This cutoff has significant ramifications for the development of stars.

2. **Nobel Prize**

Chandrasekhar's work on as far as possible procured him the Nobel Prize in Material science in 1983, over fifty years after his noteworthy exploration.

VII. Richard Feynman: Quantum Electrodynamics

American physicist Richard Feynman made critical commitments to quantum electrodynamics (QED) and acquainted novel methodologies with figuring out complex quantum peculiarities.

1. **Feynman Graphs**
   Feynman presented Feynman outlines during the 1940s, a graphical portrayal of molecule communications in quantum field hypothesis. These outlines altered the estimation of mind boggling quantum processes.

2. **Quantum Electrodynamics**

Feynman made indispensable commitments to quantum electrodynamics, giving a way to accommodate quantum mechanics with extraordinary relativity. His work added to the advancement of accuracy estimations in molecule physical science.

VIII. Murray Gell-Mann: Quarks and the Eightfold Way

American physicist Murray Gell-Mann assumed a crucial part in the improvement of the quark model and the order of hadrons.

1. **Quarks**
   During the 1960s, Gell-Mann proposed the presence of quarks, major particles that make up hadrons like protons and neutrons. His work established the groundwork for the comprehension of the solid atomic power and the characterization of particles.

2. **The Eightfold Way**

Gell-Mann presented the Eightfold Way, a grouping plan for hadrons in light of their quantum properties. This plan coordinated the plenty of particles found in molecule gas pedals.

IX. Andrei Sakharov: Sakharov Conditions

Russian physicist Andrei Sakharov is known for his commitments to molecule physical science and cosmology, especially his work on the circumstances vital for baryogenesis (the making of issue) in the early universe.

1. Sakharov Conditions

Sakharov formed the "Sakharov conditions" during the 1960s, which are three vital standards for baryogenesis to happen. These circumstances have given experiences into the early universe's way of behaving and the wealth of issue.

X. Stephen Peddling: Dark Opening Thermodynamics

English physicist Stephen Peddling made noteworthy commitments to the comprehension of dark openings, especially with his revelation of Selling radiation.

1. Selling Radiation

In 1974, Selling showed that dark openings are not altogether dark yet discharge radiation because of quantum impacts close to the occasion skyline. This revelation tested old style dark opening hypothesis and gave experiences into the interaction among gravity and quantum mechanics.

2. Commitments to Cosmology

Peddling's work stretched out to cosmology, including the detailing of the "no limit condition," which recommended that the universe had no limit or peculiarity at its starting point.

9.3 The gradual acceptance of the probabilistic nature of quantum mechanics.

Quantum mechanics, created in the mid twentieth 100 years, remains as perhaps of the best and exact hypothesis throughout the entire existence of material science. However, its probabilistic nature, described by wave capabilities and vulnerability, at first tested

the old style determinism that had overwhelmed physical science for a really long time. This exposition dives into the verifiable and logical excursion that prompted the slow acknowledgment of the probabilistic idea of quantum mechanics, featuring the key figures, tests, and philosophical discussions that molded this groundbreaking change.

1. The Emergency of Traditional Material science

To see the value in the acknowledgment of quantum mechanics, one should comprehend the emergency that traditional material science looked in the late nineteenth and mid twentieth hundreds of years.

1. **Bright Fiasco**
   The traditional hypothesis of blackbody radiation, in view of the Rayleigh-Pants regulation, anticipated an endless measure of energy in the bright locale, known as the "bright disaster." This disappointment brought up issues about the exactness of old style physical science.
2. **The Photoelectric Impact**

Albert Einstein's 1905 clarification of the photoelectric impact suggested that light comprises of discrete parcels of energy called "quanta." This thought, alongside Max Planck's work on quantization, denoted the starting points of quantum hypothesis.

II. Max Planck's Quantum Theory

Max Planck, a German physicist, assumed a urgent part in the improvement of quantum mechanics through his progressive quantum speculation.

1. **Quantization of Energy**
   In 1900, Planck presented the idea of quantization of energy, recommending that energy levels in a blackbody radiator are quantized and can exist in discrete units or "quanta." He presented the Planck consistent (h) to depict this quantization.

2. **Hesitance and Its Impact**

Planck himself was at first hesitant to embrace the ramifications of his speculation completely. Notwithstanding, his work established the groundwork for future advancements in quantum mechanics, prompting the acknowledgment of probabilistic standards.

III. Niels Bohr's Quantum Model of the Iota

Niels Bohr, a Danish physicist, took huge steps in the acknowledgment of quantum mechanics with his progressive quantum model of the hydrogen molecule.

1. **Quantization of Electron Energy**
   In 1913, Bohr's model recommended that electrons in iotas possess quantized energy levels or "shells." This model effectively made sense of the phantom lines of hydrogen, which traditional material science couldn't represent.

2. **Change Probabilities**

Bohr's model additionally presented the idea of progress probabilities, which depicted the probability of an electron changing between energy levels. This probabilistic angle tested the old style deterministic perspective on nuclear way of behaving.

IV. Wave-Molecule Duality

The revelation of wave-molecule duality further featured the probabilistic idea of quantum mechanics.

1. **Twofold Cut Investigation**
   Thomas Youthful's twofold cut explore in the mid nineteenth

century showed that light displays both wave-like and molecule like properties. This trial prepared for comparative examinations with particles like electrons, which displayed obstruction designs, affirming their wave-like way of behaving.

2. **Complementarity Guideline**

Niels Bohr's complementarity guideline, proposed during the 1920s, accentuated that particles could display both wave and molecule properties yet never all the while. This standard recognized the probabilistic idea of quantum peculiarities.

V. Werner Heisenberg's Vulnerability Rule

Werner Heisenberg's Vulnerability Rule, formed in 1927, assumed a pivotal part in the acknowledgment of quantum mechanics.

1. **Impediment of Accuracy**
   The Vulnerability Standard set that there is an essential cutoff to how unequivocally certain sets of properties, like position and force, can be all the while known. This constraint tested the old style thought of definite determinism.

2. **Philosophical Ramifications**

Heisenberg's work brought up philosophical issues about the idea of the real world, the job of perception, and the constraints of human information. It featured the inherent probabilistic nature of quantum frameworks.

VI. Erwin Schrödinger's Wave Mechanics

Erwin Schrödinger's improvement of wave mechanics gave an elective detailing of quantum mechanics that zeroed in on wave capabilities.

1. **Schrödinger Condition**
   Schrödinger's wave condition, presented in 1926, depicted the way of behaving of quantum frameworks utilizing wave capabilities. These wave capabilities addressed the likelihood conveyances

of particles, completely embracing the probabilistic idea of quantum mechanics.

2. Proportionality with Grid Mechanics

Schrödinger's wave mechanics and Werner Heisenberg's lattice mechanics were demonstrated to be identical details of quantum mechanics, stressing the consistency of probabilistic standards across various numerical systems.

VII. Banter Between Albert Einstein and Niels Bohr

The discussion between Albert Einstein and Niels Bohr over the translation of quantum mechanics gave a stage to examining its probabilistic nature.

1. Einstein's Investigate

Einstein, alongside Boris Podolsky and Nathan Rosen, proposed the EPR oddity in 1935. This oddity tested the fulfillment and probabilistic nature of quantum mechanics by featuring the thought of "creepy activity a ways off."

2. Bohr's Reaction

Bohr answered with the standard of complementarity, contending that quantum peculiarities ought to be grasped regarding corresponding portrayals, for example, wave-like and molecule like, contingent upon the trial setting.

VIII. Test Check and Acknowledgment

Notwithstanding the philosophical discussions, trial proof kept on supporting the probabilistic idea of quantum mechanics.

1. Chime's Hypothesis

John Chime's hypothesis, created during the 1960s, gave an approach to test the expectations of quantum mechanics against neighborhood authenticity tentatively. Tests roused by Ringer's hypothesis reliably upheld quantum expectations and precluded nearby secret variable speculations.

2. **Angle Investigation**

Alain Angle's examinations during the 1980s further affirmed the infringement of Ringer imbalances, giving solid proof to quantum trap and the probabilistic idea of quantum frameworks.

IX. Quantum Understandings

Different understandings of quantum mechanics have arisen, mirroring the continuous philosophical investigation of its probabilistic nature.

1. **Copenhagen Translation**
   The Copenhagen translation, advocated by Bohr and Heisenberg, stresses the job of the eyewitness and the wave capability breakdown during estimation. It recognizes the intrinsic probabilistic nature of quantum frameworks.

2. **Many-Universes Translation**
   The Many-Universes translation, proposed by Hugh Everett III during the 1950s, recommends that each quantum occasion brings about a stretching of the universe, making a large number of equal real factors. While deterministic as it were, it offers an alternate point of view on likelihood.

3. **Pilot-Wave Hypothesis**

Pilot-wave hypothesis, supported by David Bohm during the 1950s, presents stowed away factors that decide molecule directions, offering a deterministic option in contrast to the probabilistic translation.